Subsoil Improvement

Jürgen Schmitt

Subsoil Improvement

 Springer

Jürgen Schmitt
University of Applied Sciences Darmstadt
Darmstadt, Germany

ISBN 978-3-031-89748-1   ISBN 978-3-031-89749-8 (eBook)
https://doi.org/10.1007/978-3-031-89749-8

This Springer imprint is published by the registered company Springer Nature Switzerland AG
The registered company address is: Gewerbestrasse 11, 6330 Cham, Switzerland

If disposing of this product, please recycle the paper.

Preface

The stability and serviceability of structures (ground failure, slope failure, settlement, slide safety, etc.) depend on the nature and magnitude of the forces acting and the physical properties of the ground. If the properties of the subsoil do not meet the requirements to be met, an improvement of the properties can be considered.

This book describes the most common and established methods for improving the properties of the sub soil. In addition to explaining the process, it provides the requirements for using the process and information such as the maximum depths that can be improved or the strengths that can be achieved.

In addition, the book provides a comparison of methods in the form of tables, which can be used to make a rough pre-selection of possible subsoil improvement methods.

The book is aimed primarily at students and practitioners in civil engineering, architecture and engineering geology.

I have received a great deal of support in the revision of this book. Special thanks go to Mr. Christian Brandl (Laumer GmbH & Co. CSV Bodenstabilisierung KG), Mr. Dipl.-Ing. Helmut Haß (CDM Smith Consult GmbH), Mrs. Simone Hebel (Keller Holding GmbH), Mr. Dipl.-Ing. Joachim Heisler (ECOSOIL Ost GmbH), Mr. Dr.-Ing.Ing. Heiko Huber (CDM Smith Consult GmbH), Mr. Dipl.-Ing. Olaf Kloppe (ECOSOIL Ost GmbH), Mr. Dr.-Ing. Simon Meißner (Prof. Quick und kollegen—Ingenieure und Geologen GMBH), Mr. Dr. rer. nat. Joachim Michael (Prof. Quick und kollegen—Ingenieure und Geologen GMBH), Mr. Dipl.-Ing. Benno Müller (MAX BÖGL), Prof. Dr.-Ing Wolfgang Reitmeier (University of Konstanz), Dipl.-Ing. Tobias Seemann (allcons Maschinenbau GmbH), Dipl.-Ing. Marcus Scheuerer (Max Bögl), Mr. Heiner Schuchmann (Comdrill Bohrausrüstungen GmbH) and Mr. Dipl.-Ing. Christopher Tinat (Menard GmbH), who supported me in my work with technical information and illustrative pictures.

Darmstadt, Deutschland
November 2024

Jürgen Schmitt

Contents

Chapter 1
Basics

The stability and serviceability of structures (ground failure, slope failure, settlement, sliding safety, etc.) depend on the nature and magnitude of the acting forces acting and on the physical properties of the soil. If the properties of the foundation soil do not meet the requirements, an improvement of the properties can be considered (cf. [1, 2]). Whether this is economically feasible depends on the nature of the subsoil, the possibilities for improving it and the intended load. In addition to tried and tested methods, such as soil replacement and driving displacement piles, new methods based on soil mechanics have been developed in recent years or decades with the aim of broadening the field of application and making improvements more economical.

The choice of method for improving the properties of the foundation soil depends on the following boundary conditions:

- Soil Conditions,
- Groundwater Conditions,
- Type of structure,
- Actions of the structure,
- Sensitivity of the structure to settlement,
- Effects on neighbouring buildings,
- Spatial conditions, accessibility and passability of the site,
- Availability of materials, equipment, personnel,
- Economic viability,
- Environmental impact of the process.

Table 1.1 gives an overview of the most common methods of improving the properties of the foundation soil.

© The Author(s), under exclusive license to Springer Nature Switzerland AG 2025

J. Schmitt, *Subsoil Improvement*, https://doi.org/10.1007/978-3-031-89749-8_1

Table 1.1 Overview of subsoil improvement methods

Soil replacement	
Full ground replacement	Bottom part replacement
Dry dredging Dredging Box soil replacement process	Cushion layer Geotextile-covered soil columns/ geosynthetic-covered columns Concrete columns Lime-cement piles

Improvement through compaction	
Static procedures	Dynamic procedures
Preload Preload with consolidation aids Groundwater influence Surface compaction	Vibratory compaction with deep vibrator Vibratory tamping compaction Surface compaction Dynamic intensive compaction DYNIV columns Impact compaction by blasting Impact compaction with air impulse method High energy impact compaction

Improvement through consolidation	
Displacement process	Procedure without displacing effect
Vibratory compaction with deep vibrator Vibratory compaction with airlock vibrator Grouted plug columns Concrete vibrator columns Sand compaction piles CSV method Compaction injection Tear injection	Surface consolidation with binders (lime/cement) Deep soil stabilization MIP method FMI method Reinjection Jet grouting method Soil freezing

Chapter 2
Soil Replacement

In the soil replacement procedure, the insufficiently load-bearing soil is replaced with load-bearing (non-cohesive) soil. This procedure can be used economically if a load-bearing layer is overlaid by a non-load-bearing layer of low thickness and if suitable replacement soil is available at low cost.

Forr area foundations, the non-load-bearing soil is excavated, and the replacement soil is installed and compacted. Installation and compaction are generally carried out in layers above the groundwater table. The thickness of each layers depends on the compaction equipment used and the required compaction. Usually, the thickness of the layers is between 20 and 40 cm.

Below the groundwater table, layer-by-layer compaction is not possible. In this case, the inserted soil can be compacted by deep compaction (see Sect. 3.2) after backfilling has been completed. In the case of smaller replacement masses (e.g. in the case of soft silts in gravels and sands), it is generally more economical to replace the non-load-bearing subsoil with lean concrete.

In the case of subsoil with limited load-bearing capacity and foundations on single or strip footings, the installation of a compacted sand or gravel layer can be considered. The required thickness of this layer is roughly determined by the permissible load of the existing foundation soil, the existing base normal stress and the angle of propagation of the stresses. If the weight of the fill is greater than that of the existing soil, the replacement of the soil will also result in an additional load on the foundation soil.

J. Schmitt, *Subsoil Improvement*, https://doi.org/10.1007/978-3-031-89749-8_2

2.1 Full Soil Replacement

2.1.1 Dry Dredging

In dry soil replacement, the non-load-bearing soil is removed by excavators, crawler tractors or wheel loaders and transported away by trucks. The excavation pit is dug out to the intended depth and completed for backfilling of the replacement soil. Depths of up to 4 m can be achieved. The backfill material is then filled in layers and compacted. The individual layers can be compacted by static pressure (smooth roller), by beating (soil tamper), by kneading (sheepsfoot roller) or by vibration (vibrating plate, vibrating roller) (cf. [3]).

2.1.2 Dredging

For soil replacement under water, the flushing or dredging method is used. The soil is loosened mechanically, e.g. by a bucket chain dredger or in a combination with flushing assistance, and transported away in floating containers or in pipelines. Excavation depths of up to approx. 35 m below the water surface can be achieved. Before the backfill material is placed, the bottom of the excavation must be freed from silt. Special silt suction devices are used for this purpose (cf. [3]).

2.1.3 Box Soil Replacement Method

Using a box (formwork) and a backfill material, the box soil replacement method improves and compacts the soil by vibration. The box is vibrated into the soil up to the load-bearing soil layer and the contents of the box are excavated. After excavation, the new load-bearing soil is filled in and compacted by vibration during repeated lifting and lowering of the box (cf. [4]).

2.2 Base Part Replacement

2.2.1 Cushion Layer

In the case of the cushion layer, only the uppermost part of the non-load-bearing layer is replaced. In this case, excavation or replacement is only carried out to the depth required, e.g. to reduce settlement to the necessary level. A calculation example for the settlement-reducing influence of a cushion layer can be found in [5].

2.2.2 *Geotextile-Covered Soil Columns/ Geosynthetic-Covered Columns*

In the case of geotextile-covered soil columns, which are also known as geosynthetic-covered columns, three installation methods have become established: the excavation method, the displacement method and the vibro-tamping method. This means that geotextile-covered soil columns or geosynthetic-covered columns fall into different categories of ground improvement methods (cf. Table 1.1). For clarity, all three installation methods are described in this section.

In the excavation method, the soil column is produced with the aid of casing to which a high-strength casing is attached. The casing (piping) is inserted into the ground up to the load-bearing layer by vibrating or ramming. After the load bearing soil is reached, the contents of the casing are excavated and a tensile geotextile in ring or tube form is attached to the inside of the casing. The casing is then filled with new backfill material (e.g. gravel sand) and compacted by pulling out the casing. The weight of the material causes the column to expand into the surrounding soil.

A very frequently used method is the displacement method, in which a steel pipe with a closable foot flap is sunk into the load-bearing stratum. The geosynthetic casing is hooked into the steel pipe and filled with e.g. sand, gravel or crushed stone. The steel pipe is then pulled out of the subsoil under vibration. The ring pull forces are activated by the dead weight of the backfill material.

For the production of the geosynthetic-covered columns, the vibro-tamping method is used (cf. [50] and Fig. 2.1). In this process, a sealed geosynthetic sleeve is pulled over a sluice vibrator and the material pipe. The material pipe is filled (cf. Fig. 2.2). The vibrator is then driven into the ground (cf. Fig. 2.3) or displaced by the vibrator. At the same time, the geosynthetic casing is lowered into the ground. After the intended depth has been reached, the vibrator is pulled in sections and then sunk again. In the process, the material to be added (usually sand) is compacted within the geosynthetic casing (cf. Fig. 2.4). In the process, a prestress can be created. Pulling and compacting is carried out until the column is completely filled with addition material (cf. Fig. 2.5).

A load-bearing soil column with high stiffness is created (cf. [50, 53, 54]). The soil columns transfer the loads into the load-bearing soil and ensure a secure foundation. Geotextile-sheathed soil columns can also be used in very soft soils ($c_u < 15$ kN/m^2) according to [6]. Design approaches for geotextile-sheathed soil columns are given in [7, 8].

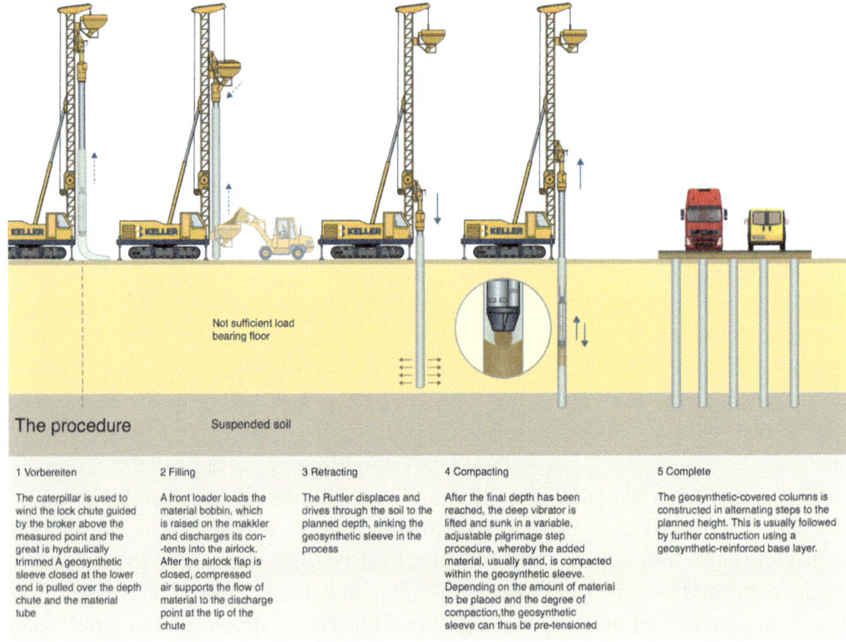

Fig. 2.1 Principle procedure geosynthetic-covered columns. (© Keller Holding GmbH)

Fig. 2.2 Filling geosynthetic covered columns. (© Keller Holding GmbH)

Fig. 2.3 Retracting geosynthetic covered columns. (© Keller Holding GmbH)

Fig. 2.4 Compacting geosynthetic covered columns. (© Keller Holding GmbH)

Fig. 2.5 Closure
geosynthetic covered
columns. (© Keller Holding
GmbH)

Chapter 3
Improvement Through Compaction

Soil compaction results in a reduction of the pore space in the soil. Compaction increases the interlocking of the soil grains and thus increases the shear strength, which ultimately increases the load-bearing capacity of the soil and reduces its susceptibility to deformation. A distinction must be made between surface and deep compaction. Surface compaction can be used to compact areas up to a depth of 2 m. Areas deeper than 2 m below the surface can only be compacted with deep compaction. Deep compaction uses only natural materials to improve the soil. Because of this, deep compaction is very environmentally friendly and can be used variably in areas of groundwater and contamination.

3.1 Static Procedures

3.1.1 Pre-loading

In order to accelerate the time course of deformations in waterlogged, fine-grained soils, which have a corresponding consolidation behaviour, preloading can be applied. Preloading displaces the pore water and the deformations can be anticipated. The uniformly applied surface loading thereby exceeds the subsequent action. The time history of the deformations depends on the constrained modulus, the permeability coefficient and the thickness of the layer to be consolidated. Further information and examples on the design of a preload can be found in [17, 18].

J. Schmitt, *Subsoil Improvement*, https://doi.org/10.1007/978-3-031-89749-8_3

3.1.2   Pre-loading with Consolidation Aids

Vertical drains can be used to accelerate consolidation in low permeability soils (see Fig. 3.1). Strip drains (cf. Fig. 3.2) consisting of a plastic core and a geotextile filter cover are frequently used for this purpose. Strip drains are installed using a stitcher (cf. Fig. 3.3). Sand/gravel drains can also be constructed. An overview of different drain materials is given in [22]. Guidance on design and dimensioning is given in DIN EN 15237. Further examples of design and dimensioning are described in [19–21, 55].

Fig. 3.1   Vertical drainage. (© MENARD GmbH)

Fig. 3.2 Strip drain. (© MENARD GmbH)

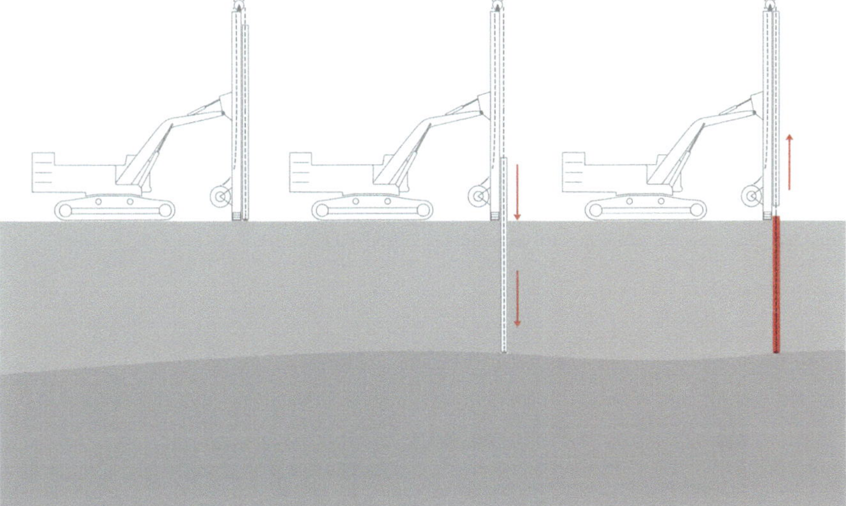

Fig. 3.3 Stitcher. (© MENARD GmbH)

3.1.3   Groundwater Influence

A preloading or a temporary acceleration of the deformations can also be achieved by a temporary lowering of the groundwater or by a vacuum consolidation. Groundwater lowering increases the effective stresses in the soil. The increase in effective stresses is effective both in the drained soil layer and in underlying soil layers. The method is particularly useful for very soft soils and soils with a very low coefficient of permeability, as only low pressure gradients can be generated in these soils using the previously described methods.

In mixed and fine-grained soils, drainage by applying a vacuum (negative pressure) is also possible. In vacuum consolidation, a negative pressure is created in the drainage elements by means of pumps (cf. Figs. 3.4 and 3.5). Examples of this are given in [23].

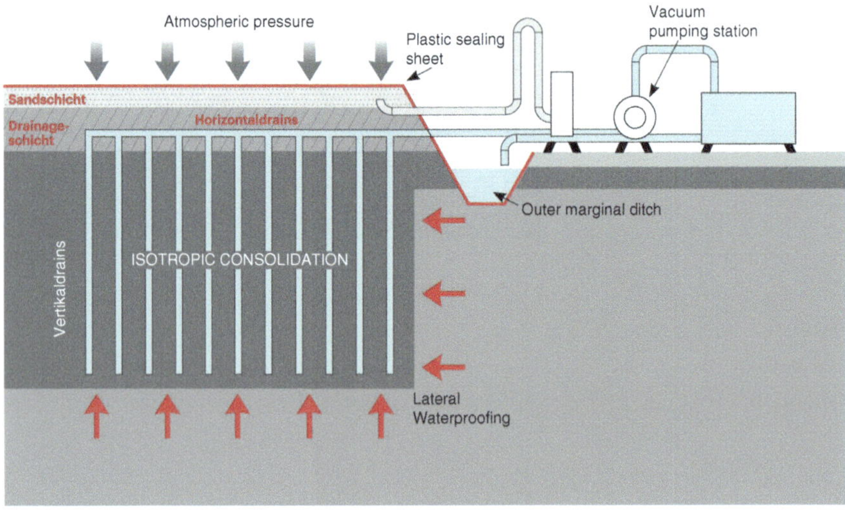

Fig. 3.4 Principle of vacuum consolidation. (© MENARD GmbH)

Fig. 3.5 Vacuum consolidation. (© MENARD GmbH)

3.1.4 Surface Compaction

Both static and dynamic compaction equipment can be used for surface compaction.

If non-cohesive soils do not have their densest bedding, they can be mechanically compacted. By reducing the pore content, the stiffness modulus and the friction angle of the soil are increased and, to a lesser extent, the permeability of the soil is also reduced. Area vibrators, vibratory rollers, vibratory tampers, explosive tampers and drop plates are used as compaction equipment. However, their depth effect is low and generally amounts to approx. 50–60 cm. Surface compaction is therefore suitable for the compaction of layers of fill. Surface compaction under the foundations is advisable if the ground has been loosened by earthworks at the level of the foundation base.

The compaction capacity of cohesive soils depends on the water content. Static-action rollers (e.g. smooth shell rollers and belt rollers) and rollers that additionally knead the soil (e.g. sheepsfoot rollers and rubber-wheeled rollers) as well as ramming equipment (e.g. drop plates and explosion rammers) are used for compaction. Area vibrators and vibratory equipment generally do not achieve good compaction with silts and clays (cf. [14]).

An overview of suitable compaction equipment depending on the material and reference values for the use of compaction equipment is presented in [24].

3.2 Dynamic Procedures

3.2.1 Vibratory Compaction with Deep Vibrator

The application of the method is intended to reduce the proportion of pores in the soil and thus increase the stiffness and shear strength of the soil. Vibratory compaction with deep vibrator (RDV) can be used to improve non-cohesive soils with a maximum of approx. 10% fine grain content (< 0.06 mm) (cf. Fig. 3.6). Due to the large proportion of pores in loosely bedded non-cohesive soils (e.g. sands and gravels), large settlements can occur when high loads are applied to the soil. In order to reduce these settlements, grain rearrangements are created in the soil by vibrations and the addition of water, which increase the density of the soil. Compaction depths of up to 25 m can be achieved.

Vibratory compaction can be carried out by means of a deep vibrator (cf. Figs. 3.7 and 3.8) or with a top-mounted vibrator.

During deep compaction with the deep vibrator, horizontal vibrations are generated in the soil with the dead weight of a vibrator and the addition of flushing water (cf. Fig. 3.9). With these vibrations, the tip of the vibrator bores up to 25 m into the soil. Depths of 65 m are possible by means of extension tubes. With the addition of a maximum of 100 m^3/h of flushing water, subsidence rates of 2–5 m per minute are achieved. Instead of flushing water, compressed air can be used as an alternative

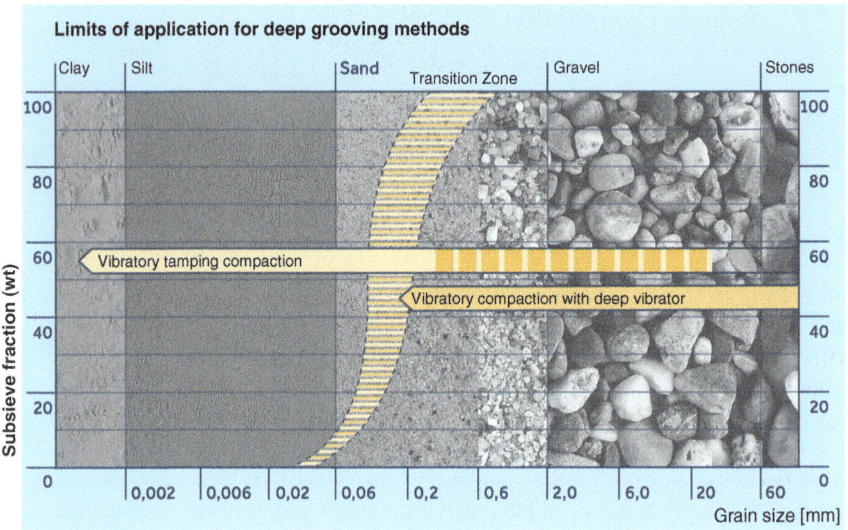

Fig. 3.6 Application limits for deep vibration methods. (© Keller Holding GmbH)

during the drilling process. After reaching the final depth, the depth vibrator is pulled out of the borehole step by step in up and down steps and the soil is compacted. The deep vibrator should be in constant contact with the soil to ensure the compaction effect. An overview of common deep vibrators is given in [25]. In order to avoid cavities (cf. Fig. 3.10), an addition material with the same or similar grain size is gradually poured into the borehole with a wheel loader. This creates a solid soil structure with a high bearing capacity in a radius of 1–2 m around the deep vibrator. This soil structure can absorb soil pressures of 800–1000 kN/m^2. The deep vibrator does not compact the soil 0.5–1 m below the ground surface. Here the soil must be compacted with surface compaction equipment. The deep vibrator has the shape of a torpedo and has a similar mode of operation as the vibrating bottle for concrete compaction. Due to the vibrations of the vibrator, the tip and the casing tube of the vibrator are heavily worn when the soil is compacted. Due to this, wear armor is applied to the tip and the casing pipe. After approximately 2000 h of operation, depending on the soil parameters, the outer casing must be replaced. Cable excavators and crawler caterpillars serve as carrier equipment for the deep vibrator. Large areas can be compacted with the aid of a ground plan grid whose points form equilateral triangles. The grid spacing should be a minimum of 1.5 m and a maximum of 3 m. The grid spacing can be determined based on empirical values. Approaches for this can be found in [15].

In deep compaction with the top-mounted vibrator, heavy top-mounted vibrators are mounted or fixed on piles, pipes, steel girders or specially shaped vibrating screeds. The piles, pipes etc. are vibrated into the ground by vertical vibrations. Here, the vibrations are generated outside and not inside as in the case of the deep

Fig. 3.7 Basic design of a deep vibrator. (© Keller Holding GmbH)

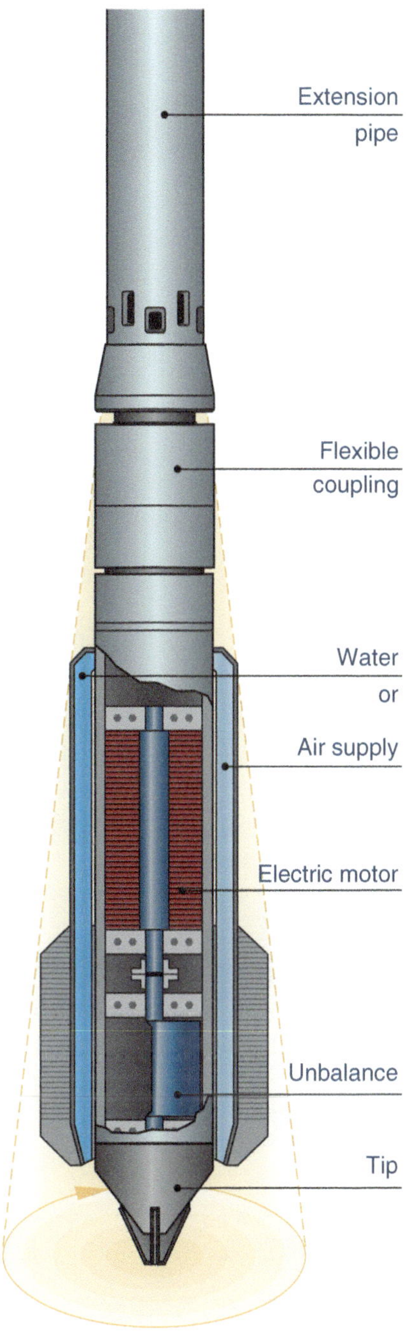

Fig. 3.8 Deep vibrator

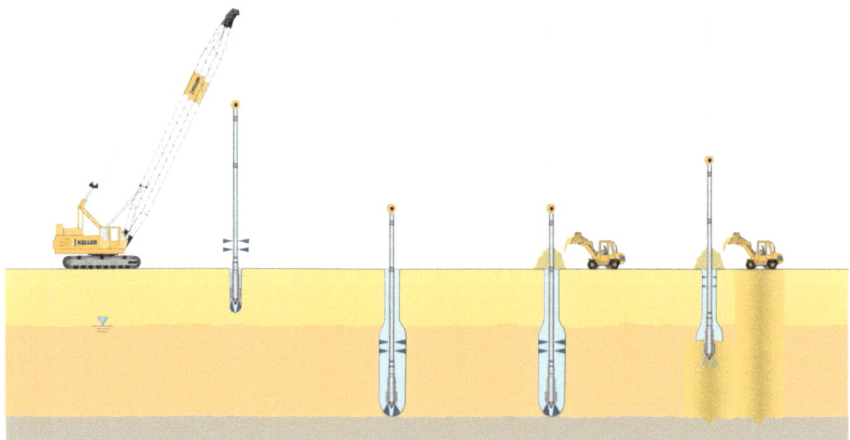

Fig. 3.9 Principle of the vibratory compaction with deep vibrator. (© Keller Holding GmbH)

Fig. 3.10 Vibrator with lowering funnel. (© Keller Holding GmbH)

vibrator. During compaction, the suspended screed is pulled out of the ground in steps, as with the deep vibrator, and compacted by adding material.

Information on dimensioning and design can be found in DIN EN 14731.

3.2.2 Dynamic Intensive Compaction

Dynamic intensive compaction (DYNIV) (cf. Fig. 3.11), which is also referred to as drop plate compaction, is mainly used for non-cohesive soils, e.g. sand or gravel above the water table. However, the method can also be used with cohesive soils e.g. silt or clay or under water, but is less effective in these applications. Compaction depths of up to 30 m can be achieved with the DYNIV. In this process, a 10–200 t mass (steel or reinforced concrete) (cf. Fig. 3.12) is dropped onto the subsoil from a height of approx. 10–40 m (cf. Fig. 3.13). In [26] a diagram can be found for estimating the effective depth of a drop plate compaction. On impact with the subsoil, shock waves are generated which cause pore water and air to escape in the soil, thus compacting the soil. This process is repeated until the soil has reached the desired bearing capacity. The number of shocks depends on the soil type, the soil depth, the drop height, the size and the impact area of the drop weight as well as the site situation. Due to the strong vibrations caused by the impact, neighbouring structures can be damaged. To avoid this, a distance to neighbouring structures of 10–50 m should be maintained. In [27] there is a diagram of the vibration effects of drop plate compaction.

The method is particularly used in road, railway and airfield construction. Due to the vibrations caused by rolling traffic, settlements form in the ground. These settlements are reduced with dynamic intensive compaction by vibration in advance.

In drop plate compaction there are three steps to improve and compact a surface or the subsoil. In this process, drop weights, which are hoisted with the aid of crawler cranes, are dropped several times from a certain height onto the grid points in a point grid. The distance between the points in the grid is approx. 4–10 m. In the first

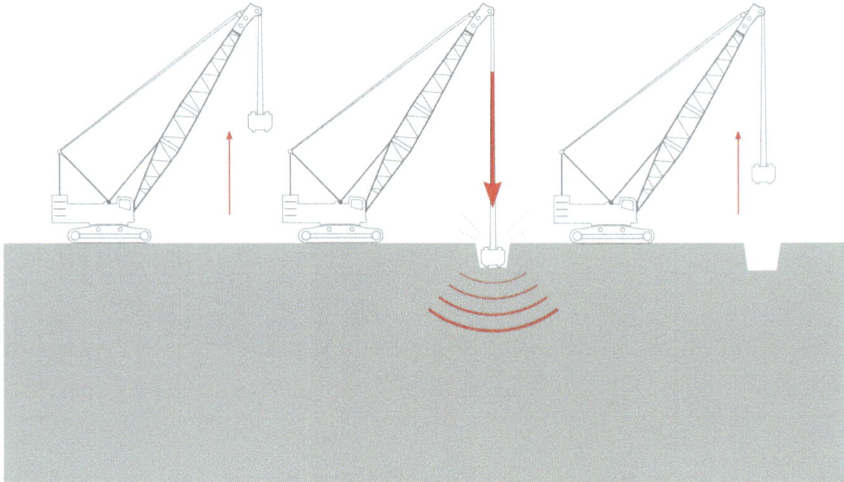

Fig. 3.11 Principle of dynamic intensive compaction. (© MENARD GmbH)

Fig. 3.12 Falling mass. (© MENARD GmbH)

step, the deepest soil layer is improved. In this process, the drop plate is dropped repeatedly at the same point from the maximum height to be reached by the crawler crane. In the second step, the drop plate is dropped between the points of the first grid from the middle height of the crawler crane. This improves the middle soil layer. In the third and final step, the near surface layer is improved with a low drop height. The impacts create compaction funnels at the ground surface (see Fig. 3.14). These funnels must be backfilled with bulk material and compacted.

Furthermore, in the case of dynamic intensive compaction, the procedure in non-cohesive soils on the one hand and the procedure in cohesive soils on the other hand must be considered. In non-cohesive soils, soil pressures of up to 700 kN/m^2 can

Fig. 3.13 Dropping the
falling mass. (© MENARD
GmbH)

Fig. 3.14 Compaction
funnel. (© MENARD
GmbH)

be absorbed after application of the method. In cohesive soils up to 400 kN/m^2 are possible.

In non-cohesive soils, the full compaction effect occurs immediately upon impact of the drop body. A firm and dense body is created in the soil, which increases the load-bearing capacity in the soil.

In cohesive soils, on the other hand, the impact of the drop body increases the pore water pressure in the soil. The excess pore water pressure can become so strong that the soil liquefies, reducing its load-bearing capacity. As a result, further compaction is no longer possible. After a resting phase in which the excess pore water pressure is reduced, compaction can be continued so that an increase in strength occurs with increasing consolidation.

3.2.3 *DYNIV Columns*

DYNIV columns are a further development of Dynamic Intensive Compaction, which is suitable e.g. for organic soils with a high water content or for heterogeneous fills. In this process, coarse-grained material is driven into the subsoil through a drop plate analogous to the DYNIV, so that a stone column is formed (cf. Fig. 3.15). The stone columns have a consolidation-accelerating effect, increase the shear strength of the soil and have a high load-bearing capacity of up to 150 t per column (cf. [65]).

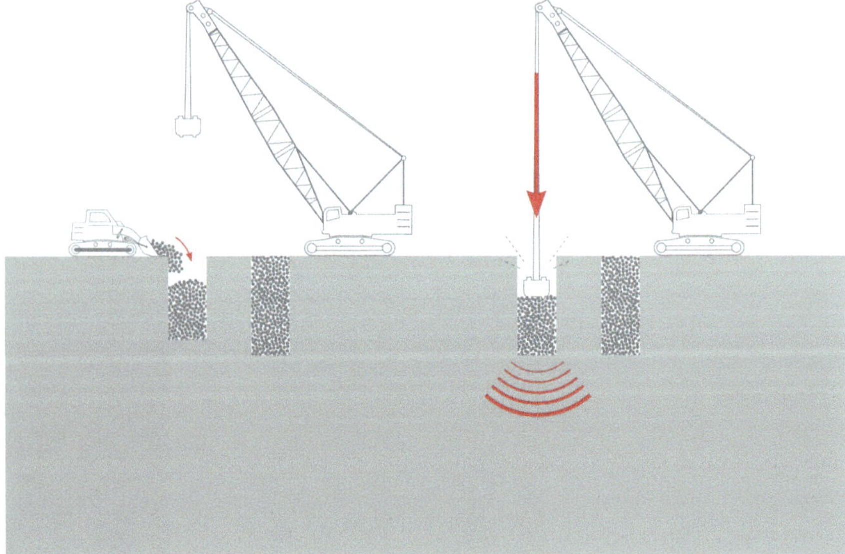

Fig. 3.15 Principle of DYNIV columns. (© MENARD GmbH)

3.2.4 Impact Compaction by Blasting

In waterlogged, loosely bedded, non-cohesive soils, compaction can be achieved by means of blasting (cf. Fig. 3.16). The aim here is to reduce the pore content of the soil in order to increase its stiffness and shear strength. A typical field of application is, for example, the rehabilitation of tipping slopes of abandoned opencast lignite mines that are at risk of subsidence.

In impact compaction by blasting, boreholes are produced using the flushing method. This is followed by the insertion and containment of the explosive charge, which is then brought to detonation. Mixtures of ammonium nitrate and TNT are frequently used as explosives (cf. [28]).

In the case of impact compaction by blasting, there is an expansion of a cavity due to expansion of the explosion vapour from the blasting. In the vicinity of the blast, i.e. in a radius of 5–10 m, compression and shear waves propagate, which lead to large plastic deformations and thus destroy the grain structure of the soil. The deformations cause excess pore water pressure. With the dissipation of the excess pore water pressure, compaction of the soil occurs (cf. [28]).

Fig. 3.16 Blast compaction. (© ECOSOIL Ost GmbH)

The blasting radius should not be less than 3 m. Recommendations for impact compaction by blasting are given in [29].

3.2.5   *Impact Compaction with Air Impulse Method*

Another method for improving settlement-prone soils is impact compaction using the air-impulse method. This combines horizontal drilling technology with the airgun technique. With the airgun technique, a defined quantity of air is released intermittently into the soil under high pressure. The escaping air mass induces a shock or impulse into the soil, resulting in cavity expansion. Further information on shock compaction with the air-pulse method is presented in [30].

3.2.6   *High Energy Impact Compaction*

High Energy Impact Compaction (HEIC) (cf. [64]) is a method for the compaction of poured or in-situ soil with an effective depth of up to 7 m. The soil is compacted by repeated roller passes at relatively high speeds. By repeated roller passes at a relatively high speed of about 12–15 km/h, the soil is improved in its bearing density and thus in its stiffness. In addition to improving the bearing density and stiffness of the soil, weak points in the subsoil and unsuitable soil inclusions are localised in the same way as with a test roller. These must either be subjected to additional treatment or excavated and filled with suitable filling materials.

Applications of high-energy impact compaction (cf. [64]) include (Fig. 3.17):

- Comparative moderation of the substrate,

Fig. 3.17   HEIC. (© LANDPAC Deutschland GmbH)

- Soil compaction of cuts, dam fills, subsoil,
- Compaction of contaminated sites,
- Compaction of cohesive, mixed-grain and coarse-grain soil types,
- Concrete asphalt crushing.

Chapter 4
Improvement Through Consolidation

In soil consolidation, hydraulic binders (cement, lime, hydraulic lime, hydrated lime) are added to the soil to protect it against stresses from traffic and the climate. The admixture permanently increases the load-bearing capacity, water insensitivity and frost resistance of the soil.

4.1 Procedures with a Displacing Effect

4.1.1 Vibratory Tamping Compaction

Due to the cohesion, cohesive soils cannot be relocated by vibration pulses as with vibratory tamping compaction. Cohesive soil types with more than 10–15% fine grain content (<0.06 mm), e.g. silts and clays, can be compacted using vibratory tamping compaction (RSV). Vibratory tamping compaction creates voids in the soil which are backfilled with gravel or crushed stone (see Fig. 4.1). This creates columns of crushed stone or gravel (cf. Fig. 4.2), which give the soil high strength and reduce settlement from building loads. The columns have a high load-bearing capacity due to the shear strength of the soil at the sides. Soil pressures of 150–400 kN/m^2 can be accommodated. The columns can be used up to a depth of 20 m and are suitable for transferring individual and area loads. They can also be used to secure structures and against ground and slope failure. For vibro-compaction, mainly tracked caterpillars are used.

Vibratory tamping compaction can be carried out with a deep vibrator (see Sect. 3.2.1 or Fig. 3.8). Alternatively, an airlock vibrator (see Fig. 4.3) can be used. Depth compaction with the airlock vibrator is similar to the procedure for vibro-compaction. A torpedo-shaped deep vibrator is vibrated into the soil up to the load-bearing soil layer using air or water flushing and then compacted by adding material such as sand, gravel or crushed stone. In contrast to vibratory compaction,

© The Author(s), under exclusive license to Springer Nature Switzerland AG 2025

J. Schmitt, *Subsoil Improvement*, https://doi.org/10.1007/978-3-031-89749-8_4

Fig. 4.1 Vibratory tamping compaction principle. (© Keller Holding GmbH)

the added material is fed into a collecting container located outside the soil at the sluice vibrator and guided to the top of the sluice vibrator by a material conveying pipe (cf. Fig. 4.4). Through the material conveying pipe attached to the side, the added material is displaced laterally into the soil using compressed air and pressed into the cavities. The resulting ballast columns have diameters of 0.6–1.0 m. As the sluice vibrator is raised, the added material is introduced in layers and compacted as it is lowered (cf. Fig. 4.5). The process is repeated until the surface is reached. The surface must be compacted with surface compaction equipment at the end, as with vibro-compaction.

For the determination of the final settlements as well as the shear parameters of the improved foundation soil, the method according to Priebe has become established (cf. [16, 48], Figs. 4.6, 4.7).

Information on dimensioning and design can be found in DIN EN 14731.

4.1.2 *Grouted Tamping Columns*

During production with a deep vibrator (cf. Fig. 4.8), a cement supension is added as a binder or a special concrete of grade C8/10 to C25/30 (cf. [31]). After hardening, the result is consolidated columns (cf. Fig. 4.9) with an average diameter of 40 cm.

Grouted tamping columns or concrete tamping columns or ready-mixed mortar tamping columns are to be designed as pile-like foundation elements. The permissible loads are up to approx. 900 kN. For the formation of buffer zones (deformation compatibility) the lower or upper part of the column can remain unconsolidated. Such columns are referred to as partially grouted tamping columns (cf. [31]).

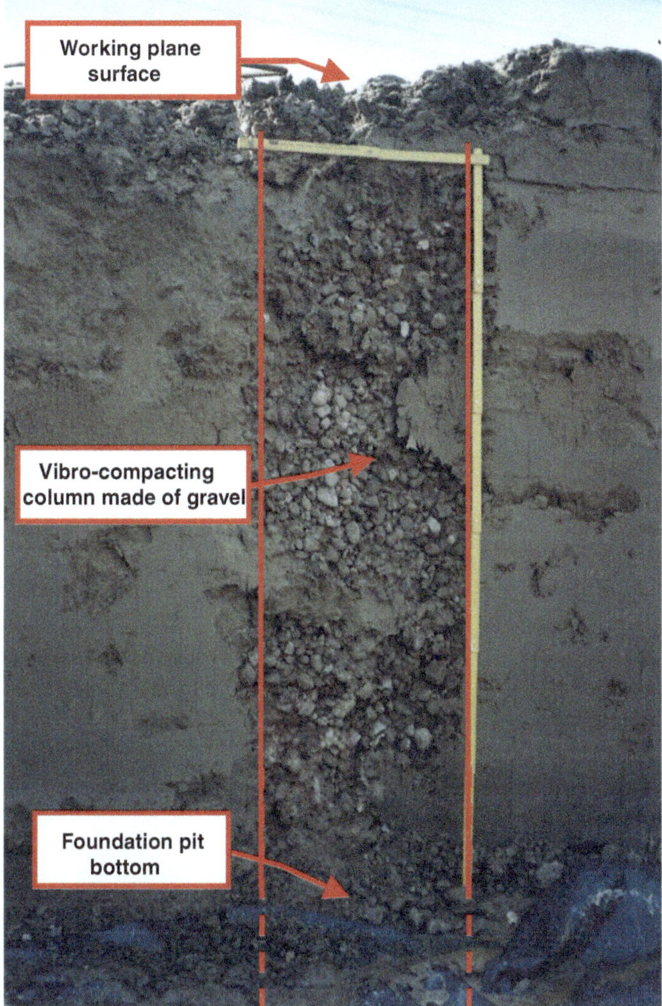

Fig. 4.2 Vibratory tamping column in a loess

4.1.3 Concrete Vibrated Columns

Concrete vibrated columns are made in their entire length from pumpable concrete of grade C 20/25 (cf. Figs. 4.10 and 4.11). Like the mortared tamping columns, concrete vibrated columns are to be designed as pile-like foundation elements. The diameter of the columns is between 40 cm and 60 cm.

The foot load capacity can be improved by lifting and lowering the vibrator several times. With a possible footing design, the load can be up to 1200 kN depending on the foundation soil (cf. [31]).

Fig. 4.3 Lock vibrator. (© Keller Holding GmbH)

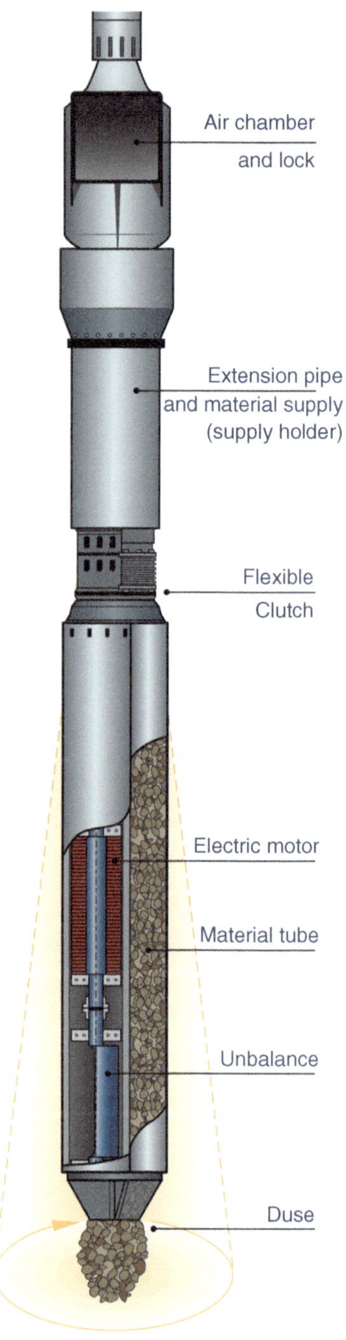

Fig. 4.4 Filling with material. (© PROF. QUICK UND KOLLEGEN – INGENIEURE UND GEOLOGEN GMBH)

4.1.4 Sand Compaction Piles

Sand compaction piles are used in soft cohesive soils as reinforcement of the soil. The sand compaction piles act as vertical drains. An open pipe is sunk into the ground under compressed air using a top vibrator and backfilled with sand, which is squeezed out when the pipe is pulled. Pipe diameters range from 0.4 to 0.6 m. In offshore applications, the pipe can have diameters up to 1.6 m. The sand compaction piles can be produced to a depth of up to 20 m. The ratio of column cross-section to grid area is 0.3–0.5 and can be up to 0.8 in the offshore area.

Reference values for the improvement in sand compaction piles are presented in [32].

4.1.5 CSV Method

In the Coplan stabilisation method (CSV), displacement columns are produced in the soil with the aid of a screw conveyor, which improve the bearing capacity of the soil (cf. Fig. 4.12). Soils that can be displaced without dynamic action are particularly suitable for this method. CSV can be used to improve non-bearing cohesive and non-cohesive soils to depths of up to 25 m. The CSV is mainly used to stabilise embankments, road, railway and flood embankments and to stabilise floor slabs and individual columns. The columns are placed by means of a point grid under planar structures. The working plane consists of a capillary-breaking layer overlying

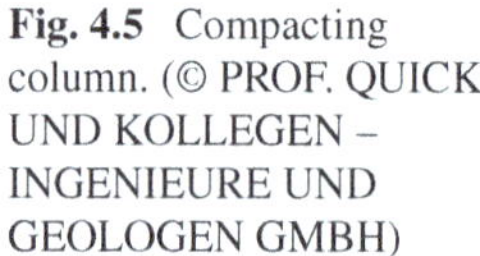

Fig. 4.5 Compacting
column. (© PROF. QUICK
UND KOLLEGEN –
INGENIEURE UND
GEOLOGEN GMBH)

a geotextile. Due to the vibration-free drainage, the displacement columns can be
constructed next to existing structures without any problems.

For the production of the displacement columns, the screw conveyor is drilled
into the ground in rotation until it reaches the load-bearing layer and a dry granulate
(lime-sand or cement-sand mixture) is added. The screw conveyor is attached to a
crawler excavator and has a receiving hopper at the lower end. The cement-sand
mixture is placed in the receiving hopper with the aid of a wheel loader and fed to the
screw conveyor. As the soil is drilled in, it is displaced laterally and compacted (see
Fig. 4.13). In the process, the dry cement-sand mixture extracts the soil moisture from
the surrounding soil. After drilling in and out, the surface is levelled and compacted
with surface compaction equipment.

Fig. 4.6 Settlement determination for individual foundations. (© Keller Holding GmbH)

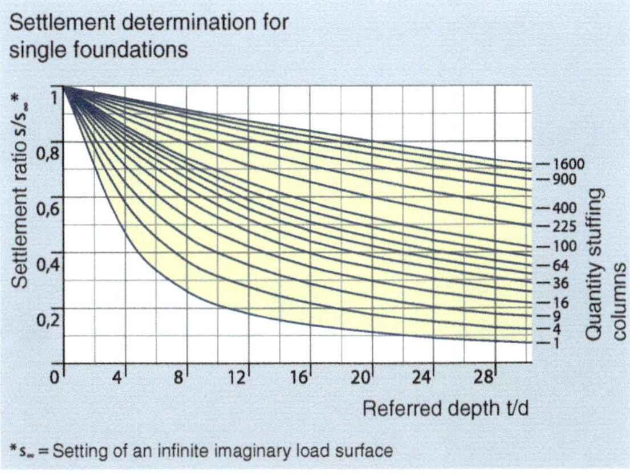

Fig. 4.7 Dimensioning diagram for vibro-compaction. (© Keller Holding GmbH)

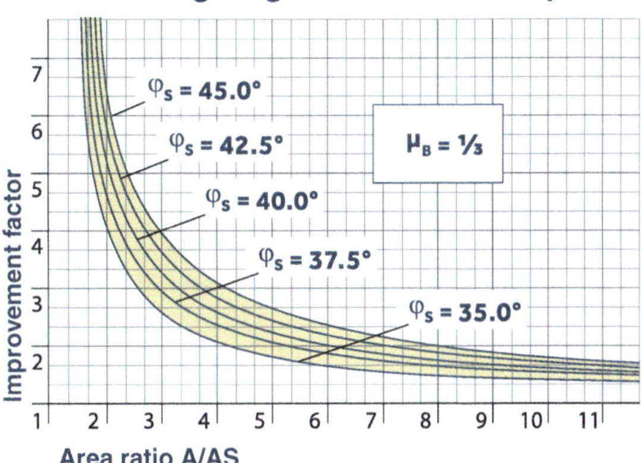

The minimum column diameter required for a secure column formation depends on the consistency of the surrounding soil. In contrast to stiff soils, where the displacement process provides a sufficiently stable pipe wall for the transport of the dry mortar, in soft and mushy soils a stable pipe wall must be created by an adapted excess of material to enable material transport without mixing with soil down to the final depth of the CSV columns. This is to be ensured in the design by pendulum steps. When setting up the unit, a check is made by pulling the transport auger without rotation to see whether there are any soil inclusions within the screw conveyor helical flights. If this is the case, the pendulum steps must be adjusted in such a way that no soil is mixed into the dry mortar. Only by observing these design rules can it be assumed that a column of dry mortar with the required minimum column diameter will be produced (cf. [49]). For soft and mushy soils, the displacement-controlled test load should be preferred (cf. [59]).

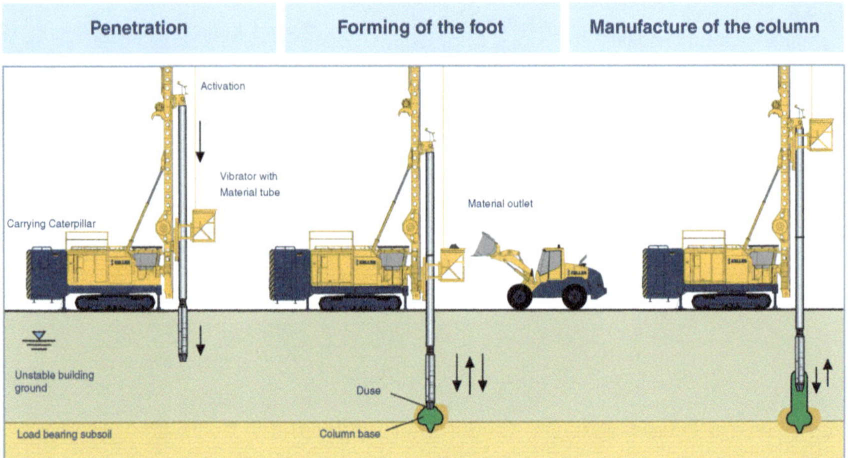

Fig. 4.8 Principle of the concrete tamping column/ready-mixed mortar tamping column method. (© Keller Holding GmbH)

With a column diameter of 16 cm, the columns can absorb soil pressures of up to 450 kN/m^2. The service load is usually between 60 and 120 kN per CSV column, depending on the soil conditions (cf. [58]). The column diameters depend on the soil conditions and range between 12 cm and 17 cm. In very soft soils, column diameters of up to 20 cm can be produced. The column spacing should be between a factor of 3 to 8 of the column diameter (cf. [10]). Column lengths of up to 9.5 are possible (cf. [58]). Between 40 and 100 m of CSV column can be produced per hour. Information on the design and design examples are given in [9, 10].

4.2 Procedures Without Displacing Effect

4.2.1 Surface Consolidation with Binders

Surface consolidation with binder can be used for cohesive and non-cohesive soils. This method is used in the construction of traffic areas and other surfaces. A distinction is made between the construction mixing method (mixed-in-place) and the central mixing method (mixed-in-plant). The difference between the two methods is that in the mixed-in-place method the soil is mixed with a binder on site and in the central mixing method the soil is mixed with a binder in a mixing plant. In the construction mixing process, the soil is torn up, crushed and mixed with the binder using a soil tiller. The binder is distributed evenly over the ground beforehand. The soil must be sufficiently moistened to ensure a chemical reaction with the binding agent. The rotation of the rotary cultivator causes the soil to mix with

Fig. 4.9 Excavated concrete tamping column. (© Keller Holding GmbH)

the binding agent at a depth of 0.5–1 m. A soil-binder mixture is produced. A soil-binder mixture is produced. This soil-binder mixture increases the shear strength and bearing capacity in the soil. After the milling operation has been completed, the soil must be compacted using surface compaction equipment, e.g. a single-drum compactor.

In the construction mixing process, the soil can be stabilised with limes or with cements. When stabilising the soil with lime, a spreader is used to spread lime on the ground (see Fig. 4.14). Additional watering of the soil is not necessary. The soil is then mixed with the binder using a rotary hoe (see Fig. 4.15). After mixing, the ground must be levelled and compacted.

When stabilizing soil with cements, addition water must still be added. There are two different ways to mix the cement with the addition water. In the first option, the cement is applied to the ground using a spreader and the water is added using a blasting truck and then mixed using a soil tiller. Alternatively, the admixture water

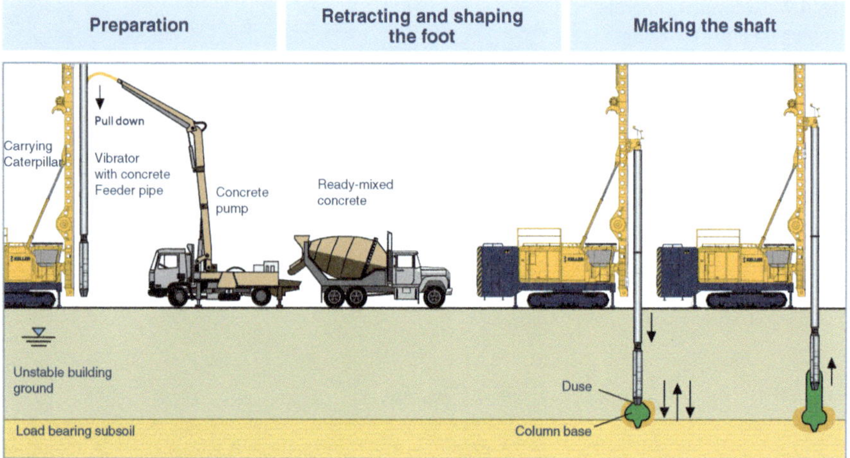

Fig. 4.10 Principle of the concrete vibrated column method. (© Keller Holding GmbH)

and cement are pre-mixed in a slurry mixer and fed to the soil tiller via a hose line. During the mixing process, the precisely metered cement suspension is then sprayed into the mixing chamber and the soil improved. As with soil stabilization with limes, the soil must finally be leveled and compacted.

Soils improved with the binder can absorb soil pressures up to 6000 kN/m^2 (cf. [12]).

4.2.2   Deep Soil Stabilisation

Deep soil stabilization was developed in Sweden and Japan and is known there as the Deep Mixing Method (DMM). Based on this, further methods were developed, e.g. Cutter Soil Mixing (CSM) or Cement Deep Mixing (CDM). An overview of the various methods can be found in [28].

In deep soil stabilization, cement, lime or other binding agents are mixed into the soil. Deep soil stabilization is standardized in DIN EN 14769. A differentiation is made between the dry-mix process and the wet-mix process.

As a representative example, this section will demonstrate deep soil stabilization for the dry mixing method using the Scandinavian method (LCM) and for wet mixing methods using the Deep Soil Mixing (DSM) method (cf. [66, 57]).

In the dry-mixing process, the mixing/agitating tool is rotated into the soil to the required depth. When the rotating boom is then pulled, the injection and mixing of the binder into the subsoil takes place through the compressed air line opening into the boom or mixing tool (cf. Figs. 4.16 and 4.17). Typical column diameters are between 60 cm and 80 cm. Depths of up to 25 m are possible. As a rule, plastic clays and silts are consolidated with lime or cement-lime mixtures as binding agents.

Fig. 4.11 Excavated
concrete vibrated column. (©
Keller Holding GmbH)

In the wet-mixing method, the soil and suspension are mixed mechanically using mixing tools. The mixing tool consists of drill pipe, paddle and drill head (cf. Fig. 4.18). During drilling, a cement suspension emerges from the nozzles located on the drill head and paddle (cf. Fig. 4.19). After reaching the required drilling depth, the rotating drill pipe is pulled and the soil is mixed with the cement suspension (cf. Fig. 4.21). The column diameter varies between 40 and 240 cm (cf. Fig. 4.20 and [66]).

In contrast to the wet-mix process, the dry-mix process is only possible in soils with a minimum water content of 20% in order to achieve a chemical reaction between the binder and the soil.

Fig. 4.12 Exposed CSV column. (© Laumer GmbH & Co. CSV-Bodenstabilisierung KG)

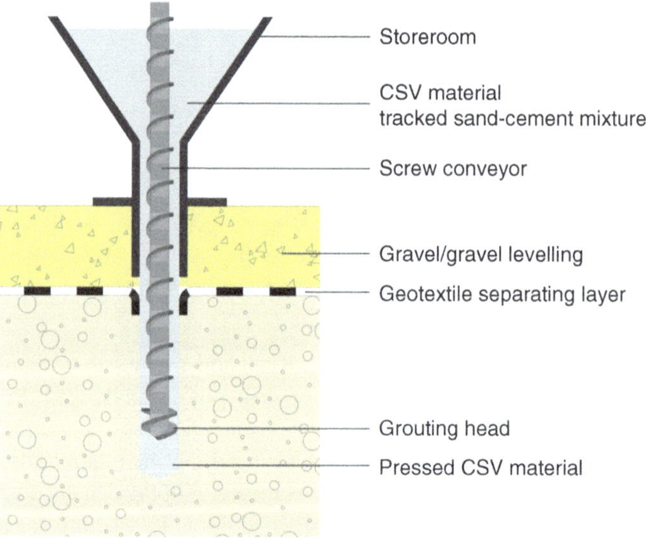

Fig. 4.13 Production of a CSV column. (© Laumer GmbH & Co. CSV-Bodenstabilisierung KG)

4.2.3 MIP Method

Another method that belongs to the group of deep soil stabilisation methods (cf. Sect. 4.2.2) is the mixed-in-place method (MIP method). With the MIP method (cf. [67]), an unsound soil is loosened by a single or triple screw conveyor that is rotated

Fig. 4.14 Spreader trolley

Fig. 4.15 Milling machine

into the soil and mixed with a cement suspension. The cement slurry is passed through the cavity of the screw conveyor to the base and used as the screw conveyor is lowered. After reaching the final depth, the soil is completely mixed and homogenized by the counter-rotating screw conveyor. A soil-suspension mixture is created in the effective area of the screw conveyor. After setting, this soil-suspension mixture forms an earth concrete body that can have a compressive strength of $300 - 1200$ kN/m^2 (cf. [12]). The MIP process is particularly suitable for non-cohesive soil, as this is itself used as a building material and can be mixed well with the cement slurry. Gravel, sand, coarse silt and stones up to about the size of a fist are well suited for mixing.

The MIP method can be used for the construction of excavation pit enclosures, for foundation measures and in flood protection. Depths of up to 25 m can be achieved (see [13]).

The production is carried out in a double pilger step process. To achieve a continuous wall, several passes are made with the screw conveyor. First, two primary

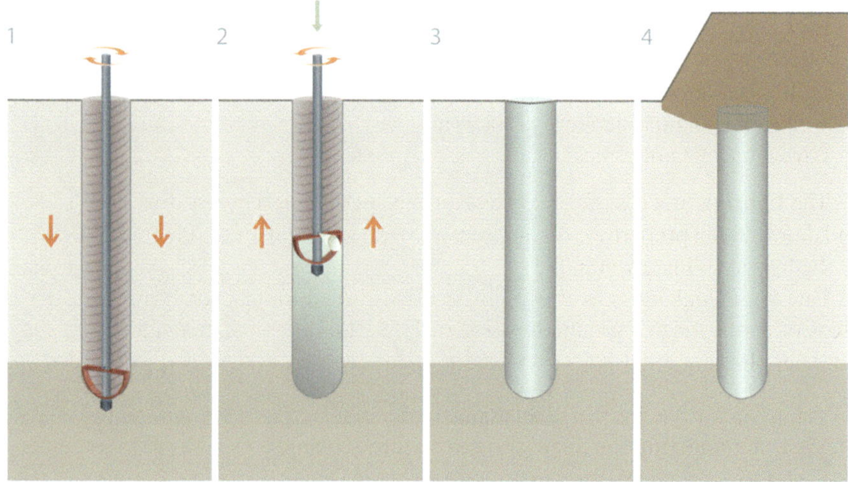

Fig. 4.16 Dry mixing process (LCM). (© Keller Holding GmbH)

Fig. 4.17 Blowing the
binder out of the mixing tool.
(© Keller Holding GmbH)

stitches are made side by side with the triple screw conveyor. Then a secondary cut
is made in the overlapping area, which is completed again by two additional cuts.
In this way, all wall sections are completely mixed and a high-quality closed wall is
created. So that the walls can take on a static function, reinforcement elements are
installed in the slots.

Information on dimensioning and design can be found in DIN EN 14679 and DIN
4093.

Fig. 4.18 Mixing tool. (© Keller Holding GmbH)

Fig. 4.19 Exit of the suspension. (© Keller Holding GmbH)

Fig. 4.20 Exposed cut-off wall with overcut DSM columns. (© Keller Holding GmbH)

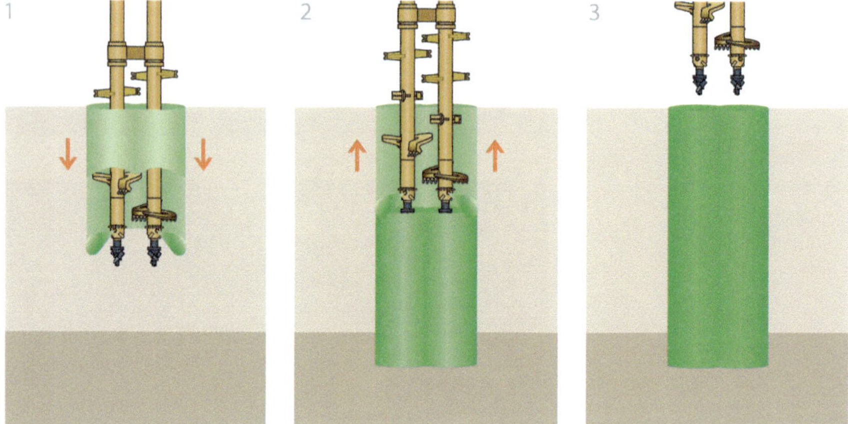

Fig. 4.21   Wet mixing process (DSM). (© Keller Holding GmbH)

4.2.4   FMI Method

The milling-mixing-injection (FMI) method is suitable for greater depths and is similar to the construction mixing method (cf. Sect. 4.2.1). The method is used for the production of diaphragm and cut-off walls, for deep soil stabilisation and for foundations. With this method, depths of up to 9 m (cf. [11]) could generally be achieved and almost all soil types improved. In the meantime, depths of up to 25 m can be achieved (cf. [62]). The minimum depth is 2 m. The slots can be produced in widths of 0.5 m and 1 m. For depths up to 9 m, the slot width is 0.5 m (cf. [11]). The tiller is larger and longer than for surface consolidation and is mounted on a tracked undercarriage (cf. Figs. 4.22 and 4.23). In a single operation, the rotating blades and the discharge of the slurry at the base of the milling blade mix and improve the

Fig. 4.22 FMI depth stabilizer. (© allcons Maschinenbau GmbH)

existing soil (cf. Fig. 4.24). For milling depths of 12 m and more, the sword is pulled out vertically up to halfway in the first step. In the second step, the milling boom is tilted hydraulically and deposited on the roof of the stabilising tiller.

Cement is the preferred binder. Depending on the type of soil, the cement content can be up to 20% by weight. The result is a jointless, weather-resistant and dense earth concrete body. One advantage of this method is that the soil material does not have to be excavated and thus no excavation support is required.

The FMI method has the general approval No. 21/97/04 of the Federal Railway Authority and is described in DIN EN 14679.

Examples of strengths achieved at 9 m milling depth are listed in Table 4.1. The performance varies depending on the milling depth between 85 m^3 per hour (2.5 m milling depth) and 800 m^3 per hour (25 m milling depth) (cf. [63]).

Fig. 4.23 FMI depth stabilizer during milling and mixing of the subsoil. (© allcons Maschinenbau GmbH)

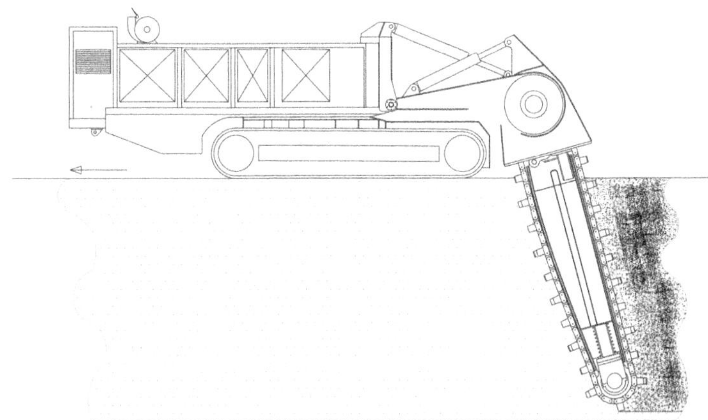

Fig. 4.24 FMI process workflow. (© allcons Maschinenbau GmbH)

Table 4.1 Unconfined compressive strengths FMI method, 9 m milling depth (according to Fa. allcons Maschinenbau GmbH [63])

Building plot	Consistency of the soils before remediation	Unconfined compressive strength after rehabilitation with FMI
Organogenic layers and sand	Pulpy, soft	$250\ kN/m^2$
Boulder clay	Soft	$750\ kN/m^2$ (after 5 days)
Weathered soils	Soft to stiff	$750\ kN/m^2$

Chapter 5
Injections

Injections are understood to be the injection (introduction under pressure) of injection agents into the cavities of the subsoil for the purpose of sealing or consolidating the subsoil. A distinction is made between filling, ripping, compacting and jet grouting.

Guidance on the design and execution of grouting is given in DIN EN 12715, DIN SPEC 18187, DIN 4093 and DIN 18309.

5.1 Basics

Cavities are defined as pores in unconsolidated rock, fissures, cracks, pores and cavernous structures in rock and in solid clay soils, and generally the contact joint between the structure and the subsoil.

The grout is a pumpable substance for filling the cavities. It must have such a small particle size that it can penetrate and fill the cavities without being filtered out.

Injection agents include cement suspensions including fine binders, mixtures of clay, water and binders, and chemical substances including plastics and bitumen (cf. [33]). If silicate gels or synthetic resins used can affect the groundwater, protective measures (e.g. pumping out or shielding the groundwater) may be necessary and must be agreed with the approval authorities.

Essential boundary conditions for injections are:

- Flow and solidification behaviour of injection agents,
- Injection pressure,
- Range,
- Injection time,
- Application areas of the individual injection methods,
- Environmental compatibility, especially of chemical injection agents.

© The Author(s), under exclusive license to Springer Nature Switzerland AG 2025

J. Schmitt, *Subsoil Improvement*, https://doi.org/10.1007/978-3-031-89749-8_5

The flow behaviour of injection agents depends on the dynamic viscosity and the yield point. The dynamic viscosity defines the penetration capacity into pores and fissures and thus determines the range of the injection agent. The yield point or initial shear strength of the viscous mass characterizes the required pressure that must be added to the actual injection pressure.

The injection pressure decisively determines how deep the injection agent penetrates into the pore or fissure system. The injection pressure depends on the one hand on the injection method and on the other hand on the material properties of the injection agent or the substrate to be injected.

For estimating the range of injections, calculation approaches are given in [35, 36].

The injection time indicates how long the injection agent is pumpable and depends on the viscosity development of the injection agent. For cement suspensions, the injection time is approx. 2 h to approx. 4 h. For conventional chemical mixtures, the injection time is reduced to approx. 20 min to approx. 60 min (cf. [34]).

Due to the rapid increase in viscosity and the available injection time, the achievable range for chemical injections is limited. The average range for solid rock is approx. 2 m to 5 m and for unconsolidated rock approx. 1–3 m (cf. [34]).

5.2 Injectables

The oldest and most common grouts are suspensions of water and cement with and without fillers (cf. Fig. 5.1). Clays, silicate gels and synthetic resins can also serve as a basis for the grout. According to DIN EN 12715, cement-based suspensions and mortars are to be used for compaction injections. Table 5.1 lists grouting agents according to DIN EN 12715 for various applications in unconsolidated and solid rock.

Fig. 5.1 Cement injection of exposed specimen. (© Dr.-Ing. Heiko Huber)

Table 5.1 Fields of application of injection agents in unconsolidated and solid rock (according to DIN EN 12715)

Subsoil	Permeability coefficient k_f/crack width e	Injectables	Type of injection
Loose rock: gravel, coarse sand, sandy gravel	$k_f > 5 \times 10^{-3}$ m/s	Pure cement suspension, cement-based suspension	Fill injection
		Cement-based suspension, mortar	Compaction injection
Loose rock: sand	5×10^{-5} m/s $< k_f < 5 \times 10^{-3}$ m/s	Fine binder suspensions, solutions	Fill injection
		Cement-based suspension, mortar	Compaction injection
Loose rock: medium to fine sand	5×10^{-6} m/s $< k_f < 5 \times 10^{-4}$ m/s	Fine binder suspensions, special chemical products	Fill injection
		Cement-based suspension, mortar	Compaction injection
Solid rock: faults, fractures, karst	e > 100 mm	Cement-based suspension, cement-based mortar, polyurethane foams, other water-reactive products	Fill injection
Solid rock: fissures, cracks	0.1 mm < e < 100 mm	Cement-based suspensions, fine binder suspensions	Fill injection
Solid rock: microfractures	e < 0.1 mm	Fine binder suspensions, silica gel, special chemical products	Fill injection
Cavities	–	Cement-based suspension, cement-based mortar, polyurethane foams, other water-reactive products	Fill injection

The flowability of the cement suspension is essentially influenced by the ratio of water to binder (water-binder value = W/B value). If the ratio value is chosen too large, cement suspensions with W/B $\geq$ 5, there is a risk that the cement particles will settle prematurely. Furthermore, a great deal of excess water must be squeezed out (filtration) before the injection is completed in order for the cement to set. For this reason, attempts are made to limit the W/B ratio to the range 0.5–1 and to improve the flowability by adding liquefying and stabilizing agents. By using special mixers, a good mixing of the solid particles with the water is enforced. Admixtures are bentonite, clay and concrete admixtures. In addition to stabilized and unstabilized cement suspensions, cement suspensions with fine-grained fillers such as fly ash, rock flour and clay are injected.

Injectables with ultra-fine binders are cements with a high grinding fineness (cf. [47]). These are obtained by individual grinding and separation of the finest constituents of the purely mineral raw materials (e.g. Portland cement clinker, granulated cinder slag and setting agent). They are mixed with water at water binder values of 0.5–8.0 by intensive mixing with a high-speed colloidal mixer to form a suspension ready for injection.

To seal the subsoil, injections of mixtures of clay, silt, sand and binders are used, especially in unconsolidated rocks. Since the clay can be washed out by erosion, a binder is added to give the clay suspension a sufficiently high and stable strength (cf. [61]). Cement is usually used, more rarely silicate-based binders. The cement also promotes coagulation (flocculation) of the clay during setting. With silicate binders, flocculation is achieved by an additional reagent.

The chemical injectables are real solutions. Their penetration capacity is very large. They consist of several components that react with each other. In the process, the solidifying or the sealing substances are formed. If the components react very quickly or "abruptly", they must be injected separately and only mix in the soil. If the reaction is slow, they can be mixed beforehand and then injected. The distance between the injection points depends on the penetration capacity and the reaction time of the injection agent used. Depending on the type of chemical substance, a distinction is made between water glass-based and plastic-based injection agents.

Silicate gel injections are based on water glass (sodium silicate $Na_2 SiO_3$). When reacting with salts, esters and aldehydes, the solution gels. Depending on the properties of the gel formed, the subsoil is mainly consolidated or sealed. If the subsoil is consolidated, the permeability is also reduced by two powers of ten. Water glass with an inorganic hardener (e.g. sodium aluminate $NaAl(OH)_4$) or organic hardener (e.g. various esters) is used for gel formation.

With a hard gel, the permeability of sands can be reduced to $k_f = 10^{-8}$ m/s to $k_f = 10^{-6}$ m/s. The unconfined compressive strength here ranges from $q_u = 1.5$ MN/m^2 to $q_u = 5$ MN/m^2. A typical formulation of an organic hard gel consists of water glass 50 vol %, water 30 vol % and hardener 20 vol %. For a soft gel, the permeability of sands can be reduced to $k_f = 10^{-8}$ m/s to $k_f = 10^{-6}$ m/s. The unconfined compressive strengths give values from $q_u = 0.3$ MN/m^2 to $q_u = 0.5$ MN/m^2. The typical formulation of an organic soft gel is composed of water glass 30 vol %, water 66 vol % and hardener 4 vol %.

In the environment of injections, depending on the composition of the silicate gel, chloride, organic carbons, sodium, calcium, magnesium, silicon, aluminum, sulfates can be detected. Changes in pH and electrical conductivity can be observed. The strongest influence on the groundwater is during the injection period. With increasing distance from the injection body and with time the influence decreases. The intensity and duration of the impact can be greatly reduced by pumping out the groundwater during the injection period. The intensity of the contamination with

Table 5.2 Resins as injection agent (according to DIN EN 12715)

Resin type	Building plot	Field of application
Acrylic resin	Granular loose rock/fine fissured rock	Reduction of permeability/increase of strength
Polyurethane resin	Large cavities	Foaming to prevent water penetration (water-reactive resins)/consolidation or local cavity filling (two-component resins)
Phenolic resin	Fine sand/sandy gravel	Sealing/consolidation
Epoxy resin	Fissured rock	Reduction of permeability/increase of strength

certain substances is superimposed by the original chemism of the groundwater and the soil.

The range of application of chemical solutions is not limited by fine particles that can adhere to the entrance of narrow pore channels or constrictions of fissures and thus prevent the penetration of further injection material. The lower application limit is given solely by the permeability of the ground to be injected and the time-dependent flow properties of the injection agent.

For injections, the water-soluble synthetic resins are of particular importance. Synthetic resins are organic substances. Synthetic resin injection agents are generally mixed before grouting. The reaction time can be largely controlled by adding catalysts. During the change of state, macromolecules are formed which cause the permanent sealing or consolidation. They are formed from the aqueous solutions by polycondensation (formation of macromolecules with splitting off of water) or by polymerization (formation of macromolecules by addition without splitting off). The synthetic resin injection agents generally have a high penetration capacity. However, they are more expensive than the chemical injection agents based on silicate or silicic acid.

Table 5.2 lists various types of resin depending on the subsoil and the possible areas of use/application according to DIN EN 12715.

Figure 5.2 shows a comparison of the application limits of injection methods and injection agents.

5.2.1 Backfilling Injections

Filling injections are used to fill pore spaces in unconsolidated rock (e.g. production of sealing blocks), fissures in solid rock (e.g. production of sealing curtains) and cavities (e.g. karst areas).

During the injection process in pore spaces, the injection agent can spread evenly radially in a homogeneous rolling soil. In a stratified soil, the injection agent spreads most strongly in the more permeable layer, so that a cylindrical body is formed

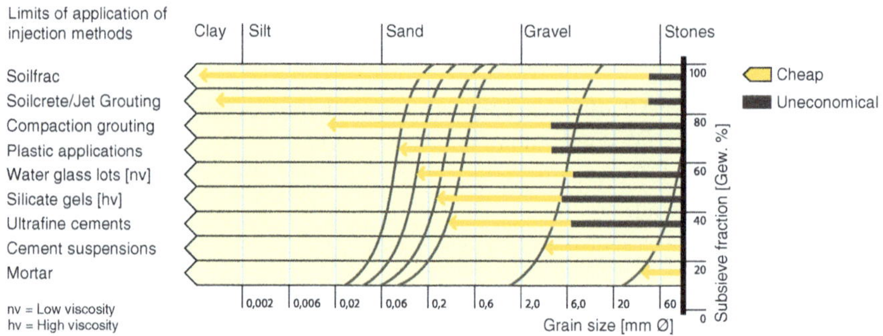

Fig. 5.2 Comparison of application limits of injections. (© Keller Holding GmbH)

there. In alternating layers with layers of different thickness and permeability, injection bodies of different shapes are therefore formed. The smaller the differences in permeability of the individual layers, the more uniformly an injection body can be produced. Therefore, it has to be clarified for the injection process whether the intended range of the injection is larger than the thickness of the layer to be injected or smaller. Depending on this, the spacing of the injection elements laterally and vertically in the sequence of layers must be determined. It is advisable that the most permeable layers are injected first to reduce the permeability differences of the remaining layers. The injectability of sands with silt content is difficult to estimate. Even silt contents of 5–10% can prevent injection.

In the case of pore injection, a flushing hole is first made. This is filled with a barrier agent. Then a valve tube is adjusted. After the sealing agent has hardened, the valve opening (rubber sleeve) is approached with the double packer. The double packer is fixed in the valve tube and the valve opening is pressed open. After the barrier agent has burst open, injection takes place in the pore space. After the pore space has been injected, the double packer is moved to the next valve opening.

A detailed explanation of pore space injection can be found in [34].

Injections into fractures and fissures in solid rock are usually aimed at sealing the rock mass. A characteristic feature of injection into fissured rock is that the rock bodies between the fractures (interfaces) are generally dense and solid and water often flows in the fractures at considerable velocities. The water squeeze (WD) test is used to assess the injectivity of fractures, despite considerable interpretive difficulties. Since fractures are often intersected by fractures, the flow conditions change with increasing distance from the borehole. The flow process is determined by the fracture shapes, fracture width, fracture roughness, fracture directions and any fracture fills.

When injecting in solid rock, the boreholes are usually injected in defined sections. The size of the individual sections is usually several metres. As a rule, several fissures are captured and injected at the same time. The individual borehole sections are closed off at the top and bottom by means of packers. Packers with hydraulically or pneumatically expanding sealing collars are most frequently used. With mechanical

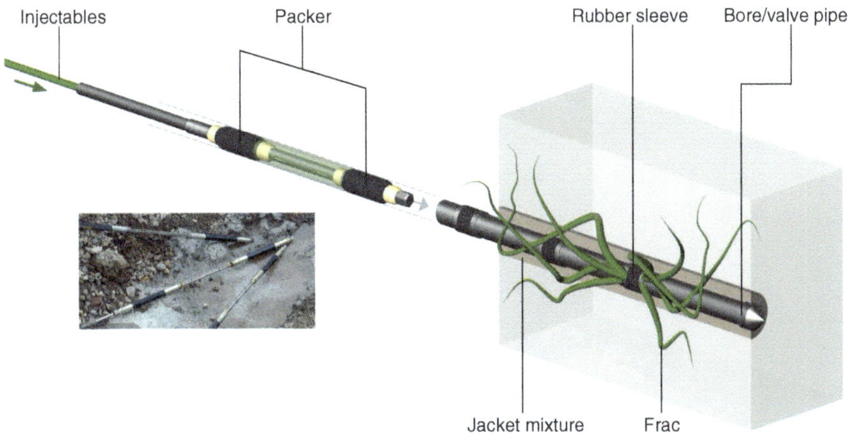

Fig. 5.3 Schematic of Soilfrac process. (© Keller Holding GmbH)

packers, the sealing sleeve is pressed against the borehole wall with a screwing device.

5.3 Tear Injections

Tear injections (Soilfrac® method, cf. [42, 52]) are intended to artificially create new flow paths and fill existing cavities, fissures and pores with injection agent (cf. Fig. 5.3). Multiple grouting at the same injection point leads to increased frac formation or rupturing. With continued systematic injection, uplift can be achieved as a result.

Cement suspensions with or without mineral additives are used as injection agents.

In non-cohesive soils, injection can also be carried out in a short sequence during heave. In cohesive soils, the excess pore water pressure must first subside before re-injection can take place.

Tear injections are used, for example, in foundation rehabilitation or, for example, to compensate for settlements caused by heave injections during tunnel driving (cf. Fig. 5.4).

5.4 Compaction Injections

In compaction grouting, a stiff to plastic grout is usually injected into the soil under pressure (cf. Fig. 5.5). The ground is displaced or compacted in the immediate vicinity of the injection source. The grout remains as a coherent body. The grout does not significantly penetrate the soil pores, nor is the soil split open. Non-load bearing soil

Fig. 5.4 Soilfrac method Installation of horizontal valve pipes. (© Keller Holding GmbH)

areas below solid soil layers can be reached and improved. Fissures and cavities in the soil can be filled and compacted under pressure and thus in a force-fit manner.

A similar degree of improvement in the subsoil can be achieved by compaction injection as with a deep vibration method. Compaction injection is particularly suitable for improving subsoil where space is very limited or only a limited working height is available. Compaction injection can also be used as a vibration-free method, e.g. due to sensitive neighbouring buildings or if compaction is to be carried out at a particularly great depth (cf. [43]).

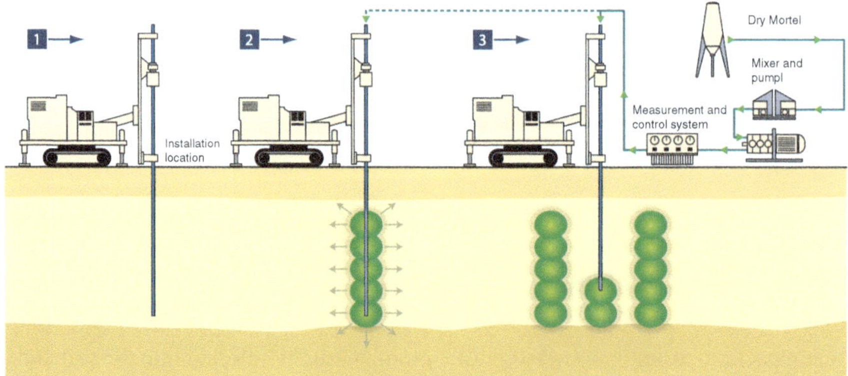

Fig. 5.5 Compaction injection method. (© Keller Holding GmbH)

5.5   Jet Grouting Method

In the conventional methods described in Sects. 5.3, 5.4 and 5.5, existing pore spaces, fissures and possibly cavities are filled with the injection agent. The structure of the soil remains practically unchanged.

In the jet grouting process, the subsoil is milled (eroded) by high-energy jet streams so that it can mix with the injected suspension. The jet grouting method is therefore not an injection method in the conventional sense.

Jet grouting is a process also known as jet-grouting, soilcrete and high pressure injection (HDI) (see [44, 45]).

Guidance on the design and execution of jet grouting is given in DIN EN 12716, DIN 18321 and DIN 4093.

As a rule, the drill pipe is lowered with the injection nozzles installed by means of rotary drilling and external flushing (max. depth in Germany 27 m, in Japan up to 50 m). After reaching the final depth, the fluid flow is switched from drilling mud to nozzle injection and the suspension emerges through the nozzles under high pressure of up to 800 bar (and at high speed of at least 100 m/s). The soil is milled up in the area of the nozzle jet. By rotating the boom during the drawing process, a column-shaped injection body is formed. The excess amount of suspension (mixed with fine particles of soil) rises in the annular space of the borehole and escapes uncontrolled at the upper edge of the borehole or it is discharged via a relief borehole.

Usually, the excess quantity of the suspension is pumped into tight settling basins or containers and, after hardening, recycled, e.g. in earthworks or road construction. Screening of coarse fractions, regeneration of the binder content and reuse is only successful in exceptional cases. In polluted soils or groundwater, the excess quantity of the suspension brings pollutants to light, which must be disposed of properly.

Further technical data (reference values) for the nozzle jet process are nozzle number 1 to 2, nozzle diameter 1.5–4.5 mm, drawing speed 4–80 cm/min, number of revolutions 3–10 rpm.

Depending on the arrangement of the injection points, single columns (cf. Fig. 5.6), closed walls (cf. Fig. 5.7) or interlocked cuboid bodies can be produced. If the rod is pulled without rotary movement, thinner wall discs are produced (sealing wall).

There are two different high-pressure media and three process variants for nozzle jetting. Cement or water can be used as the high-pressure medium. The process variants available for selection are pure suspension jet (single process/S = Single, cf. Fig. 5.8), air-jacketed suspension jet (double process/D = Double, cf. Fig. 5.9) and air-jacketed high-pressure water jet and below suspension jet (triple process/T = Triple, cf. Fig. 5.10).

The jet grouting method can basically be used in all loose rock soils. In the case of silts and fine sands, there is homogeneous mixing of the soil and groundwater with the binder suspension and a homogeneous soil mortar is formed. In coarse sands and gravels, there is homogeneous mixing with subsequent sedimentation effects in the cubature and a soil mortar with a variable proportion of "aggregates" over the height

Fig. 5.6 Exposed Soilcrete
body underpinning in sandy
soil. (© Keller Holding
GmbH)

Fig. 5.7 Exposed Soilcrete
wall. (© Keller Holding
GmbH)

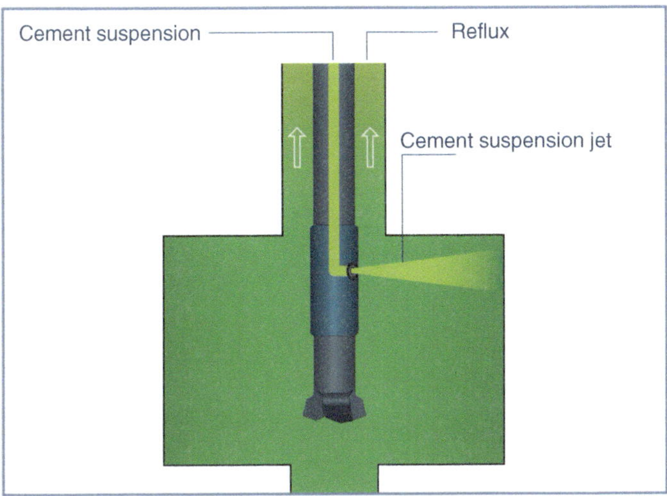

Fig. 5.8 Process variant Soilcrete-S. (© Keller Holding GmbH)

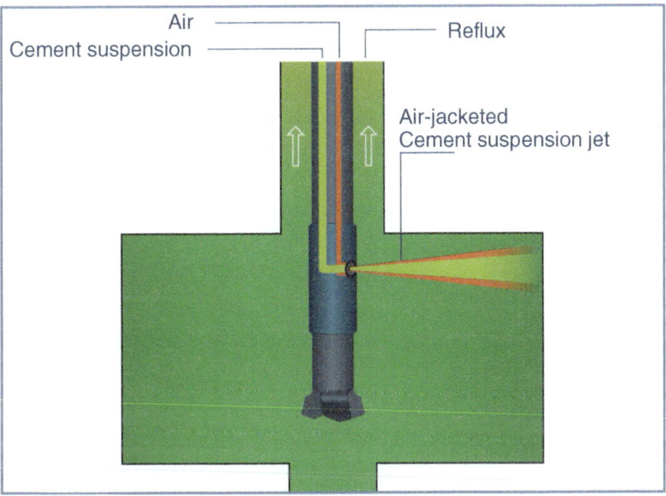

Fig. 5.9 Process variant Soilcrete-D. (© Keller Holding GmbH)

results. In plastic clays, suspension and soil separate in cubature and the column consists largely of cement paste.

A major advantage of the jet grouting method is that building materials are used which exclude any risk to groundwater, as the suspensions contain only mineral components (cement, bentonite, sand). The cement content is between approx. 200–300 kg per m^3 cubature in the case of rolling soils, between approx. 250–350 kg per m^3 cubature in the case of weakly cohesive, silty soils, and between approx.

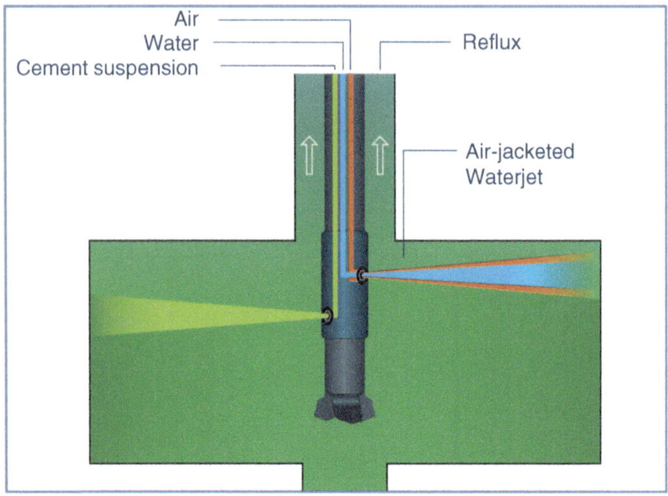

Fig. 5.10 Process variant Soilcrete-T. (© Keller Holding GmbH)

300–400 kg per m^3 cubature (and more) in the case of cohesive, clayey soils, and in the case of strongly clayey soils there is extensive replacement.

The decisive parameters for the planning of jet grouting are the achievable column diameter, the achievable strength and the achievable permeability.

The diameter of the columns is approximately 0.4–0.5 m (1.0 m) in clayey silt, 0.8–0.9 m (1.6 m) in silty sand, and 0.9–1.0 m (2.4 m) in sandy gravel. The values in parentheses apply to the Soilcrete-T method. This method requires special measures such as an air-jacketed nozzle jet.

The achievable strengths are 10 MN/m^2 to 25 MN/m^2 in sand and gravel, MN/m^2 in silt 5 (cf. Fig. 5.11) and less than 3 MN/m^2 in organic soils.

The water permeability of intact, homogeneous jet stream cubatures is in the range of $k_s = 10^{-8}$ m/s to $k_s = 10^{-12}$ m/s, depending on the soil fraction. The system permeability achieved in-situ is almost completely determined by the proportion of local or systematically occurring leakage points.

The diameter of the columns can be determined by core drilling obliquely through the columns, screen measurements (not a standard procedure), ultrasound and similar procedures in the fresh column, uncovering of sample columns as well as by the Acoustic Column Inspector ACI® (see [46]). Since the density of the water-soil-suspension mixture cannot be measured with sufficient accuracy (missing reference section), the ultrasonic method is still very inaccurate. Exposing sample columns may not be possible in some cases due to depth or confined space conditions. The Acoustic Column Inspector (ACI) is a method that uses piezo sensors to register the vibrations generated by the nozzle jet when it hits level bars. Thus, the range can be determined acoustically as well as optically.

One of the main applications for the jet grouting method is underpinning, as underpinning using the jet grouting method is exceptionally low in deformation in

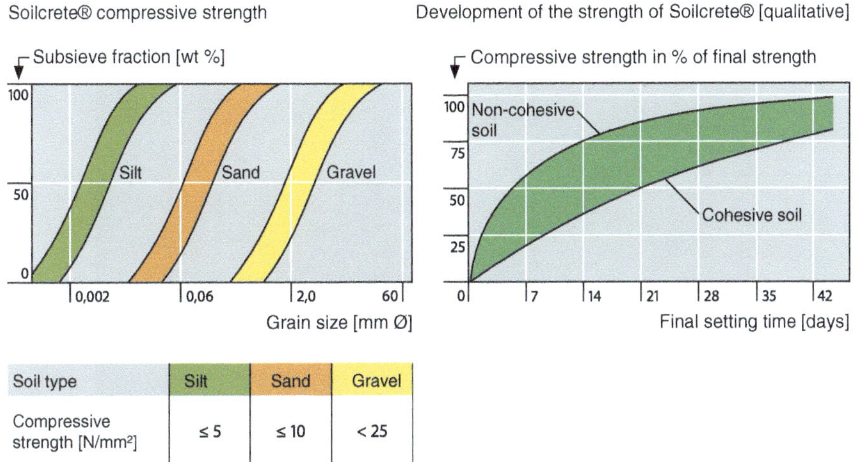

Soil type	Silt	Sand	Gravel
Compressive strength [N/mm²]	≤ 5	≤ 10	< 25

Fig. 5.11   Soilcrete method compressive strengths. (© Keller Holding GmbH)

the final state. During production, deformations can occur due to the temporary loss of foundation surfaces and due to procedural errors. For this reason, production is carried out using the pilgrim step method. Underpinnings should consist of at least two rows of columns to allow for safety margins. The jet grouting method is also used for the rehabilitation of foundations that are no longer sufficiently stable, e.g. in historic buildings (cf. Fig. 5.12). Other areas of application (cf. [51, 56]) are sealing bases, sealing blocks for the exit and entrance of tunnel boring machines, ridge protections for tunnels in loose rock, outer shells for the construction of tunnel cross passages.

Fig. 5.12 Soilcrete column underpinning historic building. (© Keller Holding GmbH)

Chapter 6
Soil Freezing

The strength of frozen soils is significantly higher compared to the unfrozen state. In addition, frozen soils are impermeable to water. The principle of soil freezing is based on the approach of artificially extracting heat from the soil and freezing it with the accompanying drop in temperature below the freezing point of water. In order to carry out soil freezing, it is necessary that sufficient water is available in the pores of the soil or in the fissures of the rock for freezing. In the case of non-water-saturated and non-cohesive soils, water can be added.

In order to initiate heat extraction in the ground, a freezing pipe is inserted into the ground. In the freezing pipe there is a downpipe which reaches to the lowest point of the pipe. At the inlet of the downpipe, the refrigerant is introduced, exits at the bottom and flows back in the annulus. In contact with the freezing tube, the refrigerant extracts heat from the adjacent soil. The heat exchange that is made possible creates a frost body (cf. Fig. 6.1).

For the production of the frost body by means of heat extraction, two processes are available: brine freezing and nitrogen freezing. Nitrogen freezing can usually be used economically for small freeze volumes and short freeze holding times. The time required to create a frost body with nitrogen freezing is approx. 2–4 times less than with brine freezing. The two processes can also be combined. Thus, in a first step, shock freezing can be performed using nitrogen freezing to create the frost body. In a second step, the more cost-effective brine freezing is then used to maintain the frost body (cf. [38]).

Ground freezing is fundamentally divided into three phases: the freezing phase, the holding phase and the thawing phase.

In the freezing phase, the frost body is completely formed and the static thickness is reached. Here, the cooling unit must provide maximum power until the process is completed. Cylindrical bodies are formed, which in the final phase form frozen walls or plates.

After the formation of the frost bodies, the cooling capacity must be adjusted to compensate for the heat quantities acting on the frost body from outside. This

J. Schmitt, *Subsoil Improvement*, https://doi.org/10.1007/978-3-031-89749-8_6

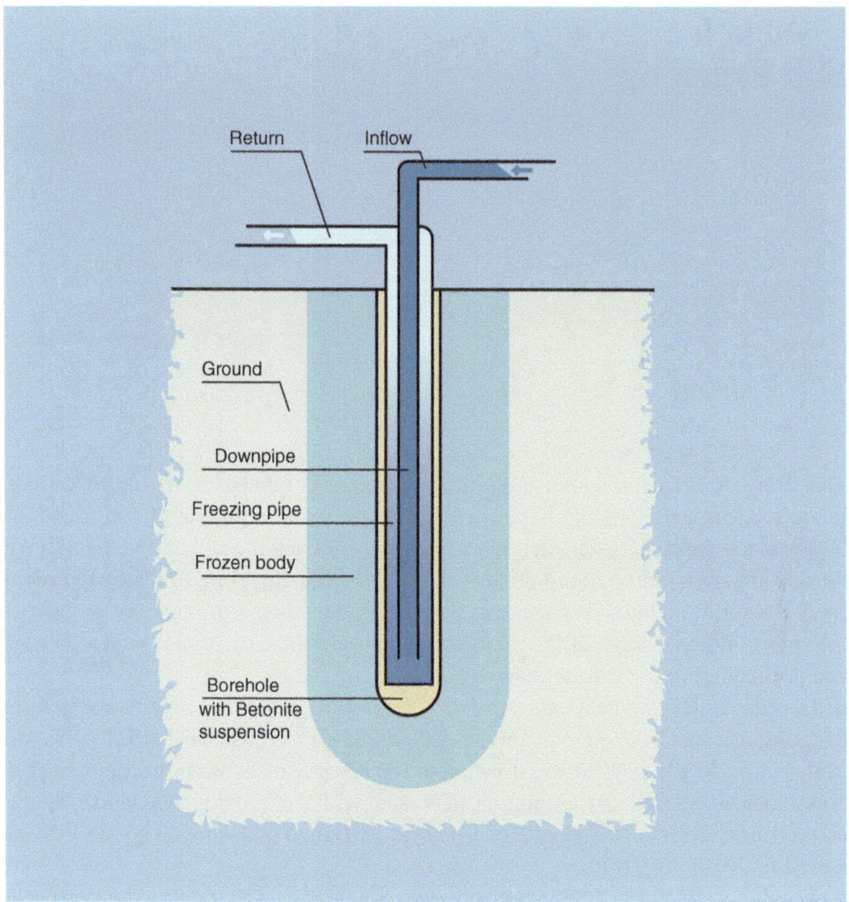

Fig. 6.1 Principle of soil freezing. (© Max Bögl Group)

phase defines the holding phase. During this phase, the freezing plant may be operated temporarily with some interruptions or with a higher inlet temperature of the refrigerant. The frost body must not lose the statically necessary strength during this phase. At the same time, the frost body should not become larger, as this can lead to complications within the construction work.

After completion of the construction measures, no further securing by ground freezing is required, so that the thawing phase takes place in the final step. All cooling units are shut down and put out of operation so that the ground can thaw naturally.

Flowing groundwater can strongly influence soil freezing. Permanent heat input can delay or prevent frost body closure (cf. [37]). As guideline values for an uncritical groundwater flow velocity, v < 2 m/d can be considered for brine icing and v < 4–6 m/ d for nitrogen icing.

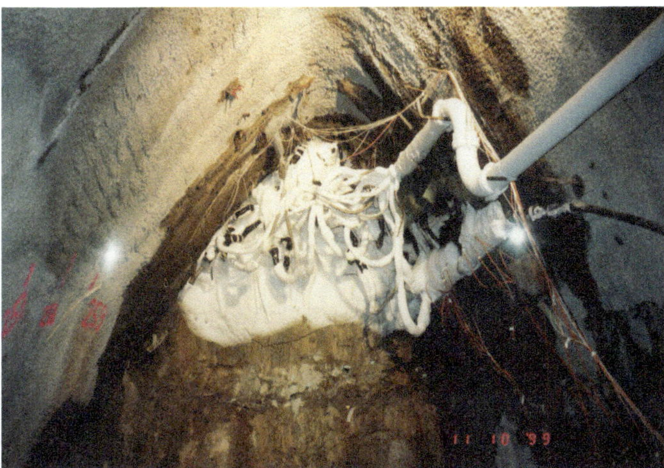

Fig. 6.2 **Soil** freezing in the elm tunnel during conventional tunnel driving. (© CDM Smith Consult GmbH)

The following can be considered as a possible additional measure to be able to control groundwater flows with soil freezing measures:

- Reduction of the distance between the freezing tubes,
- Injections to reduce permeability,
- Insertion of additional freezing pipes in the approaching area,
- Well installation to reduce groundwater gradient,
- colder freezing temperature of the brine,
- Use of nitrogen in critical areas.

The freezing process of the soil changes its thermal properties as well as strength and deformation properties. A comprehensive presentation of the changed properties of the frozen soil is given in [41].

Due to the high costs, soil freezing is usually used for temporary measures. The areas of application of ground freezing (cf. [60] and Figs. 6.2, 6.3 and 6.4) include, for example:

- Tunnelling: ridge freezing, heading support, construction of cross passages,
- Start-up protection for shield drives and pipe jacking,
- Underpinning, securing of excavation pits and shafts,
- Collection of undisturbed soil samples,
- Pressing through railway underpasses,
- Remediation.

Fig. 6.3 I cing recovery tunnel boring machine. (© CDM Smith Consult GmbH)

Fig. 6.4 Soil freezing in tunnel construction. (© Max Bögl Group)

6.1 Nitrogen Freezing

In nitrogen freezing, liquid nitrogen (LN_2) is used as a refrigerant. The liquefied nitrogen is delivered to the site at $-$ 196 °C in specially insulated tankers and temporarily stored there in tank farms. The liquid nitrogen reaches the freezing pipes via an insulated piping system. The liquid nitrogen is fed via the downpipe into the freezing pipe, evaporates and escapes in gaseous form into the atmosphere. As a result, temperatures between $-$ 60 °C and $-$ 100 °C prevail in the escaping gas (cf. [39]). In the soil, temperatures of $-$ 20 °C to $-$ 30 °C are reached, depending on the distance from the freezing pipe (cf. Figs. 6.5 and 6.6).

The very low temperatures create a large temperature gradient. This is also referred to as shock freezing. In the freezing phase, 1500–2500 l of liquid nitrogen are required

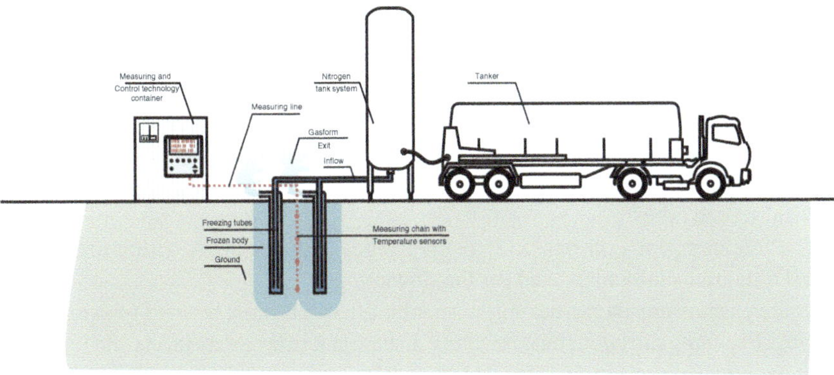

Fig. 6.5 Principle of soil freezing with nitrogen. (© Max Bögl Group)

Fig. 6.6 Nitrogen freezing. (© Dr.-Ing. Heiko Huber)

for 1 m^3 of frozen soil as a rough approximation. For the holding phase, as little as 90 l/ m^3 per day is sufficient (cf. [37]). These values vary very strongly with the thermal characteristics of the soil in place, the water content, the groundwater flow velocity and the ambient conditions. In nitrogen freezing, a temperature-controlled nitrogen metering system with solenoid valves is often installed for each freezing tube. This allows each freeze tube to have its freezing capacity controlled independently of the adjacent freeze tubes.

The freezing pipe consists of 2″ copper pipes. Copper pipes with a diameter of 15–20 mm are usually also used for the downpipes.

In the case of nitrogen freezing, particular attention must be paid to safety engineering. Escaping nitrogen must be safely discharged from excavations and tunnels in order to avoid displacement of vital oxygen. Otherwise, there is a risk of suffocation in the event of accumulation in excavation pits or tunnels.

6.2 Brine Icing

In brine icing, an aqueous salt solution is used as the cold carrier. A calcium chloride solution (CaCl$_2$) with a mass content of approx. 30% is very often used. The lowest melting point is approx. $-$ 55 °C.

In contrast to nitrogen freezing, brine freezing is a closed system in which the aqueous solution circulates in an insulated ring line and is repeatedly cooled by a cooling unit (cf. Fig. 6.7).

In the refrigeration unit, the brine solution is cooled down to the desired flow temperature using refrigerants (e.g. ammonia, carbon dioxide, fluorinated hydrocarbons). The system is composed of three heat circuits: recooling circuit, refrigerant circuit and brine circuit. In a refrigeration cycle, the gaseous refrigerant is sucked

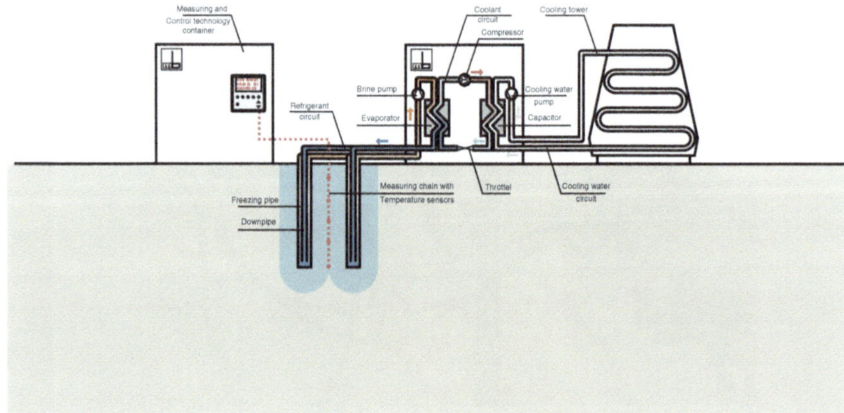

Fig. 6.7 Principle of soil freezing with brine. (© Max Bögl Group)

in and strongly compressed in a compressor. The gas, which is under high pressure, is cooled in a heat exchanger (condenser) and converted from a gaseous to a liquid state. The heat quantity during the cooling and the change of aggregate state is mostly dissipated to the environment by cooling water and a cooling tower. However, air-cooled condensers are also used. The resulting liquid refrigerant is led to a throttle element (expansion valve) and expanded. During expansion, the refrigerant pressure decreases while the liquid refrigerant cools down. In the downstream evaporator, the refrigerant changes from the liquid to the gaseous state again. The energy required by the refrigerant for evaporation (heat of evaporation) is extracted from the brine solution, which cools down to approx. $-35\,°C$ in the process. The compressor sucks in the evaporated refrigerant again and the cycle process is closed (cf. [39]). In the brine circuit, the brine loses 2–5 °C compared to the flow temperature by absorbing thermal energy from the ground.

According to Harris [40], approx. 460 W cooling capacity per metre of freezing pipe can be assumed as a reference value for the required capacity at a brine temperature of $-30\,°C$. This can be used as a basis for dimensioning a refrigeration system. This can be used as a basis for dimensioning a refrigeration system.

The average temperatures in the frost body are between -10 and $-20\,°C$ during brine icing.

The freezing pipes are made of steel with an outer diameter of 90–120 mm. The downpipe is made of polyethylene and is inserted centrally into the freezing pipe. In order to avoid high temperature losses, as with nitrogen freezing, insulation is necessary for the pipe system consisting of the supply and return pipes.

In frost-sensitive cohesive soils, there is a risk of the formation of ice lenses during freezing and the associated heaving and later settlement during thawing. In fine-grained soils, negative pressure is created at the frost line and water is sucked in. The accumulated and freezing water then leads to ice lenses and frost heave. These deformations must be considered and evaluated in both methods. Pulsating operation of the refrigeration system can inhibit the formation of ice lenses, limiting heaves.

Chapter 7
Comparison of Procedures

In Tables 7.1, 7.2, 7.3, 7.4 and 7.5, the various subsoil improvement methods are compiled by means of method comparisons. These process comparisons serve as a basis for making a rough preselection for possible ground improvement processes. These are then to be examined in detail with regard to feasibility, economic efficiency and environmental compatibility.

J. Schmitt, *Subsoil Improvement*, https://doi.org/10.1007/978-3-031-89749-8_7

Table 7.1 Comparison of methods—reference values for maximum improvable depths

Procedure	Maximum improvable depth
Full soil replacement (dry dredging)	Up to approx. 4 m
Box soil replacement process	Up to approx. 7 m
Pre-charge with consolidation aids	Up to approx. 10 m
Vibratory compaction with deep vibrator	Up to approx. 25 m
Vibratory compaction with deep vibrator with extension	Up to approx. 65 m
Surface compaction	Up to approx. 2 m
Dynamic intensive compaction	Up to approx. 10 m resp. up to approx. 30 m
High energy impact compaction	Up to approx. 7 m
Vibratory tamping compaction	Up to approx. 20 m
Sand Compaction Piles	Up to approx. 20 m
CSV method	Up to approx. 25 m
Surface consolidation with binders (lime/cement)	Up to approx. 1 m
MIP method	Up to approx. 25 m
FMI method	Uup to approx. 15 m resp. up to approx. 25 m
Injections	Unlimited
Jet grouting	Up to approx. 30 m
Soil freezing	Unlimited

Table 7.2 Comparison of methods—reference values for absorbable soil pressures

Procedure	Maximum absorbable soil pressure
Full ground replacement	Up to approx. 700 kN/m^2
Vibratory compaction with deep vibrator	Up to approx. 1000 kN/m^2
Dynamic intensive compaction	Up to approx. 400 kN/m^2 or up to approx. 700 kN/m^2
Vibratory tamping compaction	Up to approx. 400 kN/m^2
CSV method	Up to approx. 450 kN/m^2
Surface consolidation with binders (lime/cement)	Up to approx. 6000 kN/m^2
MIP method	Up to approx. 1200 kN/m^2
FMI method	Up to approx. 750 kN/m^2
Injections	Up to approx. 5000 kN/m^2 or up to approx. 20,000 kN/m^2
Jet grouting	Up to approx. 500 kN/m^2 resp. up to approx. 5000 kN/m^2
Soil freezing	Up to approx. 500 kN/m^2 or up to approx. 1500 kN/m^2

Table 7.3 Suitability

Evaluation criteria	Full ground replacement	Pre-charge with consolidation aids	Vibratory compaction with deep vibrator	Dynamic intensive compaction	Vibratory tamping compaction	CSV method	Surface comp. with binders	MIP method	FMI method	Injections	Jet grouting	Soil freezing
Application cohesive soils	+ +	+ +	−	+	+ +	+ +	+ +	+	+ +	0/ +	+	+ +
Application non-cohesive soils	+ +	−	+ +	+ +	0	+ +	+ +	+ +	+ +	+ +	+ +	+
Absorbable soil pressures	+	−	+ +	+	+	+	+ +	+ +	+ +	+ +	+ +	+ +
Increase in storage density	+ +	−	+ +	+ +	+	0	+	0	0	0/ +	−	0
Increase in load capacity	+ +	−	+ +	+ +	+	+	+ +	+ +	+ +	+ +	+ +	+ +
Improvable depth	0	0	+ +	+	+ +	+ +	−	+ +	+	+ +	+ +	+ +
Reduction of settlements	+	+ +	+	+ +	+	+	0	+	+	+	+ +	+ +

− = not suitable/0 = conditionally suitable/ + = well suitable/ + + = very well suitable

Table 7.4 Economic efficiency/boundary conditions construction site

Evaluation criteria	Full ground replacement	Pre-charge with consolidation aids	Vibratory compaction with deep vibrator	Dynamic intensive compaction	Vibratory tamping compaction	CSV method	Surface comp. with binders	MIP method	FMI method	Injections	Jet grouting	Soil freezing
Distance to neighbouring buildings	g	G	G	G	G	g	g	g	g	g	g	g
Space requirement	−	−	+	−	+	+	−	+	0	−	−	−
Equipment costs	−	−	0	0	0	0	−	0	0	−	−	−
Material requirements	−	−	0	0/ +	0	0	−	−	−	−	−	−
Personnel requirements	−	−	−	0	−	0	−	−	−	−	−	−
Groundwater lowering required	J	N	N	N	N	N	J	N	N	N	N	N

− = high/0 = medium/ + = low/G = high/g = low/Y = yes/N = no

Table 7.5 Environmental boundary conditions

Evaluation criteria	Full ground replacement	Pre-charge with consolidation aids	Vibratory compaction with deep vibrator	Dynamic intensive compaction	Vibro-tamping	CSV method	Surface comp. with binders	MIP method	FMI pmethod	Injections	Jet grouting	Soil freezing
Risk of soil contamination	N	N	N	N	N	J	J	J	J	N	N	N
Vibration	−	+	−	− −	−	+	+	+	+	+	+	+
Noise development	−	+	−	−	−	+	−	+	−	+	+	+
Dust development	−	+	+	−	+	+	−	+	+	+	+	+

+ = low/ − = high/ − − = very high/Y = yes/N = no

Chapter 8
Standards and Recommendations

DIN EN 12715:2000-10: Execution of special geotechnical works—Grouting.

DIN EN 12716:2019-03: Execution of special geotechnical work—Jet grouting method; German version EN 12716:2018.

DIN EN 14679:2005-07: Execution of special geotechnical works—Deep mixing.

DIN EN 14679:2006-09: Execution of special geotechnical works—Deep mixing, Corrigenda to DIN EN 14679:2005-07.

DIN EN 14731:2005-12: Execution of special geotechnical works—Ground treatment by deep vibration.

DIN EN 15237:2007-06: Execution of special geotechnical works—Vertical drainage.

DIN SPEC 18187:2015-08: Supplementary provisions to DIN EN 12715:2000-10, Execution of special geotechnical works—Grouting.

DIN 4093:2015-11: Design of strengthened soil—Set up by means of jet grouting, deep mixing or grouting.

DIN 18309:2016-09: German construction contract procedures (VOB)—Part C: General technical specifications in construction contracts (ATV)—Ground treatment by grouting.

DIN 18321:2016-09: German construction contract procedures (VOB)—Part C: General technical specifications in construction contracts (ATV)—Jet grouting work.

J. Schmitt, *Subsoil Improvement*, https://doi.org/10.1007/978-3-031-89749-8_8

References

1. Lackner, E. (1996). Schwierige Gründungen in Verbindung mit Bodenverbesserungen. *Der Bauingenieur, Heft 9.*
2. Seegrön, F.-A., & Laparose, J. (1978). Erfahrungen bei der Erschließung von Wohnbauland, Industrie-und Gewerbeflächen, Straße und Autobahn (Heft 10).
3. Maybaum, G., Mieth, P., Oltmanns, W., & Vahland, R. (2011). *Verfahrenstechnik und Baubetrieb im Grund-und Spezialtiefbau* (2nd ed.). Vieweg + Teubner Verlag.
4. Fa. Möbius GmbH; Firmenprospekt: Kasten-Bodenaustausch-Verfahren.
5. Smoltczyk, U., & Hilmer, K. (1991). Baugrundverbesserung in: Grundbau-Taschenbuch, Teil 2: Geotechnische Verfahren (4. Auflage), Ernst & Sohn Verlag.
6. Kempfert, H.-G., & Raithel, M. (2015). GEOTECHNIK nach Eurocode, Band 1: Bodenmechanik (4. Auflage), Beuth Verlag.
7. Raithel, M. (1999). Zum Trag-und Verformungsverhalten von geokunststoffummantelten Sandsäulen, Heft 6, Schriftenreihe Geotechnik, Universität Kassel.
8. Raithel, M., & Kempfert, H.-G. (2000). Bemessung von geokunststoffummantelten Sandsäulen, *Die Bautechnik, 76*(Heft 12), S. 983–991.
9. Reitmeier, W. (2013). Baugrundverbesserung nach dem CSV-Verfahren, *Die Bautechnik, 90*(Heft 9), 539–549.
10. Merkblatt für die Herstellung: Bemessung und Qualitätssicherung von Stabilisierungssäulen zur Untergrundverbesserung, Teil I–CSV-Verfahren DGGT (2002).
11. Ast, W., & Polloczek, J. (2002). FRÄS-MISCH-INJEKTIONSVERFAHREN (FMI), PLANUNGSHILFE—Hinweise für den Einsatz bei der DB Netz AG, Stand 10 July 2002.
12. Prinz, H., & Strauß, R. (2011). Ingenieurgeologie (5. Auflage), Spektrum Akademischer Verlag.
13. Wehr, W. (1999). Schottersäulen—Das Verhalten von einzelnen Säulen und Säulengruppen. *Geotechnik, Heft, 1,* 40–47.
14. Voss, R., & Floß, R. (1968). Die Bodenverdichtung im Straßenbau. 5. Auflage, Düsseldorf.
15. Thorburn, S. (1975). Building structures supported by stabilised ground. *Geotechnique, 25*(1975), 83–94.
16. Priebe, H. J. (1995). Die Bemessung von Rüttelstopfverdichtungen, *Die Bautechnik, 72*(Heft 3), S. 183–191.
17. Merkblatt über Straßenbau auf wenig tragfähigem Untergrund, FGSV (2010).
18. Kamiolkowski, M., Lancellotta, R., & Wolski, W. (1983). Precompression and speeding up consolidation. *General Report, Proceedings 8th European Conference Soli Mechanics and Foundation Engineering* (Vol. 3, pp. 1201–1226).
19. Calderon, P. A., & Romana, M. (1997). Soil Improvement by precharge and prefabricated vertical drains at Tank Group No. 3, site at the TOTAL oil storage plant at valencia harbour. *14th International Conference on Soil Mechanics and Foundation Engineering*, Hamburg (Vol. 3, S. 1577–1580).

J. Schmitt, *Subsoil Improvement*, https://doi.org/10.1007/978-3-031-89749-8

20. Sievering, W. (1985). Bodenverbesserung durch Auflast und Vertikaldrainage. *Geotechnik, Heft, 3*, 115–119.
21. Mohamed, R., Zakaria, F., & Gofar, N. (2008). Performance of ground improvement by precompression and vertical drain, Geotropkia 2008, Kuala Lumpur, 26–27 May 2008.
22. Kirsch, K., & Bell, A. (2013). *Ground improvement* (3rd ed.). CRC Press.
23. Schiffer, W., Varaksin, S., & Chaumeny, J. L. (1994). Vakuumkonsolidierung von frisch aufgeschüttetem Boden am Beispiel des Ausbaus Vorwerker Hafen in Lübeck, Baugrundtagung Köln (S. 233–241).
24. Witt, K. J. (2018). Erdbau in: Grundbau-Taschenbuch, Teil 2: Geotechnische Verfahren (8. Auflage), Ernst & Sohn Verlag.
25. Kirsch, K., & Kirsch, F. (2017). *Ground improvement by deep vibratory methods* (2nd ed.). CRC Press.
26. Slocombe, B. C. (1993). Dynamic compaction. In M. P. Moseley (Ed.), *Ground improvement* (pp. 20–39). Blackie Academic & Professional.
27. Varaksin, S. (1990). Neuere Entwicklungen von Bodenverbesserungsverfahren und ihre Anwendung, 5. Christian Veder Kolloquium, Neue Entwicklungen in der Baugrundverbesserung, Institut für Bodenmechanik, Felsmechanik und Grundbau, TU Graz.
28. Witt, K. J. (2018). Baugrundverbesserung und Injektionen in: Grundbau-Taschenbuch, Teil 2: Geotechnische Verfahren (8. Auflage), Ernst & Sohn Verlag.
29. Lausitzer und Mitteldeutsche Bergbau-Verwaltungsgesellschaft mbH (MBV); Universität Karlsruhe, Technische Universität Bergakademie Freiberg. (1998): Beurteilung der Setzungsfließgefahr und Schutz von Kippen gegen Setzungsfließen, Anlagenteil, S. 3–17.
30. Tudeshki, H. H. (1997). Das Luft-Impuls-Verfahren: Darstellung der Entwicklung und der Wirkungsweise eines neuen Verfahrens zur Verdichtung lockergelagerter Böden; dargestellt am Beispiel der Sicherung und Sanierung setzungsfließgefährdeter Kippen und Kippenböschungen von Tagebauen, RWTH Aachen.
31. Fa. Keller Grundbau GmbH; Firmenprospekt: Die Tiefenrüttelverfahren, Prospekt 10–02 D.
32. Aboshi, H., Mizuno, Y., & Kuwabara, M. (1991). Present StaFte of sand compaction pile in Japan, deep foundation improvement, ASTM STP 1.089, Esrig, M. I. and Bachus. R. C., American Society for Testing and Materials.
33. Jessberger, H. L. (1970). Die Einpreßmittel für den Baugrund und ihr Verhalten bei Einpassungen. Z. VDI-Zeitschrift (1970) Heft 3.
34. Kutzner, C. (1991). Injektionen im Baugrund, Enke Verlag.
35. Müller-Kirchenbauer, H. (1969). Untersuchungen zur Eindringung von Injektionsmassen in porigen Untergrund und zur Auswertung von Probeverpressungen, Institut für Bodenmechanik und Felsmechanik, Technische Hochschule Friderciana Karlsruhe, *Heft 39*.
36. Wittke, W. (1968). Zur Reichweite von Injektionen in klüftigen Fels, Felsmechanik und Ingenieurgeologie, Suppl. IV, S79–89.
37. Sres, A. (2009): Theoretische und experimentelle Untersuchungen zur künstlichen Bodenvereisung im strömenden Grundwasser, Veröffentlichungen des Instituts für Geotechnik (IGT) der ETH Zürich, DISS ETH Nr. 18378.
38. Martak, L., Haberland, C., Wolf, W., & Weigl, H. (2005). U-Bahn Wien/A: Bergmännischer Vortrieb unter dem Donaukanal im Schutz einer Baugrundvereisung mit kombinierter Stickstoff- und Solemethode. *STUVA Forschung und Praxis, Heft, 41*, 116–122.
39. Fa. Implenia Spezialtiefbau GmbH; Firmenprospekt: Bodenvereisung.
40. Harris, J. S. (1995). *Ground freezing in practice*. Thomas Telford.
41. Witt, K. J. (2018). Bodenvereisung in: Grundbau-Taschenbuch, Teil 2: Geotechnische Verfahren (8. Auflage), Ernst & Sohn Verlag.
42. Fa. Keller Grundbau GmbH; Firmenprospekt: Das Soilfrac®-Verfahren, Prospekt 61–02D.
43. Fa. Keller Grundbau GmbH; Firmenprospekt: Die Verdichtungsinjektion, Prospekt 66–01D.
44. Fa. Keller Grundbau GmbH; Firmenprospekt: Das Soilcrete®-Verfahren, Prospekt 67–03 D.
45. Fa. BAUER Maschinen GmbH; Firmenprospekt: HDI Bauer Hochdruckinjektion.
46. Fa. Keller Grundbau GmbH; Firmenprospekt: Acoustic Column Inspector—ACI®, Prospekt/Brochure 67–04D/E.

47. Tausch, N., & Teichert, H.-D. (1990). Injektionen mit Feinstbindemitteln—Zum Eindringverhalten von Suspensionen mit Mikrodur in Lockergestein; 5. Christian-Veder-Kolloquium: Neue Entwicklungen in der Baugrundverbesserung, 26.-27.04.1990, Graz.
48. Priebe, J. (2003). Zur Bemessung von Rüttelstopfverdichtungen – Anwendung des Verfahrens bei extrem weichen Böden, bei „schwimmenden" Gründungen und beim Nachweis der Sicherheit gegen Gelände-oder Böschungsbruch. *Die Bautechnik, 80*(Heft 6), S. 380–384.
49. Reitmeier, W., Stallhofer, A., & Gaissmaier, M. (2019). Baugrundverbesserung nach dem CSV-Verfahren im Straßenbau am Beispiel der Sanierung der Bundesstraße B32 in Altshausen, Baden-Württemberg, 1. Kolloquium Straßenbau in der Praxis, *Technische Akademie Esslingen, 29*, 30 Jan 2019.
50. Fa. Keller; Firmenprospekt: Geokunststoffummantelte Säulen für den Verkehrswegebau, Prospekt 12–03 D.
51. Schrank, M., Rathmair, F., & Martak, L. (2010). 25 Jahre Düsenstrahlverfahren Fortschritt in Technologie und Anwendung, 25. Christian Veder Kolloquium, 08.–09.04.2010.
52. Otterbein, R., & Dekker, H. (2000). Anwendung des Soilfrac-Verfahrens in Europa, Internationaler Studientag "Compensation Grouting", Antwerpen, 08 Dec 2000.
53. Wehr, J., & Heerten, G. (2005). Geogitterummantelte Säulen—ein neues System der Baugrundverbesserung, 9. *Informations-und Vortragstagung für Kunststoffe in der Geotechnik, FSKGEO, München, 2005*, 223–227.
54. Sidak, N., & Strauch, G. (2003). Herstellung geotextilummantelter Kiessäulen mit Keller-Tiefenrüttler, 4. Österreichische Geotechniktagung in Wien, 24–25 Feb 2003.
55. Wehr, J., Rawitzer, S., & Mock, J. (2004). Konsolidierung organischer Böden mittels Rütteldrains—vorgestellt an zwei Bauvorhaben, 11. Darmstädter Geotechnik Kolloquium, 18 Apr 2004 (S. 129–136).
56. Stellte, M., & Weber, A. (2010). Moderne Sicherungstechniken mit Injektions-und Düsenstrahlverfahren bei aktuellen Tunnelprojekten. *BauPortal, 4*(2010), 226–232.
57. Breitsprecher, G., Schuhmacher, N. (2009). Neueste Entwicklungen und Anwendungen beim DSM- und TSM-Verfahren, 16. Darmstädter Geotechnik Kolloquium, 19 Mar 2009 (S. 67–79).
58. Fa. Laumer Bautechnik GmbH, Firmenprospekt: Spezialtiefbau, CSV-Verfahren und Mikropfähle.
59. Reitmeier, W., & Brandl, C. (2012). Vergleich von kraft-und weggesteuerten Probebelastungen an CSV-Säulen, Grundlagen, Interpretation und Erfahrungswerte, 2. Symposium Baugrundverbesserung in der Geotechnik, Technische Universität Wien, 13–14 Sep 2012 (S. 99–114).
60. Fa. MAX BÖGL, Firmenprospekt: Infrastruktur Bodengefriertechnik.
61. Cambefort, H. (1969). *Bodeninjektionstechnik—Einpressungen in Untergrund und Bauwerke*. Bauverlag.
62. Fa. allcons Maschinenbau GmbH, Firmenprospekt: products, services and solutions.
63. Fa. allcons Maschinenbau GmbH, Funktionsbeschreibung: Bindemittelgabe + Fräsen und Mischen in einem Arbeitsschritt mit dem innovativen Fräs-Misch-Injektionsverfahren (FMI), Stand: 25 July 2018.
64. Fa. ECOSOIL Ost GmbH, Firmenprospekt; HEIC High Energy Impact Compaction.
65. Fa. MENARD GmbH: Firmenprospekt: Effiziente und nachhaltige Baugrundverbesserung, Verfahrensübersicht.
66. Fa. Keller Grundbau GmbH; Firmenprospekt: Tiefreichende Bodenstabilisierung, Bodenverbesserung durch Bodenmischverfahren, Prospekt 32–01 D.
67. Fa. BAUER Spezialtiefbau GmbH; Firmenprospekt: BAUER Mixed-in-Place.